MULTIPLE-INPUT, MULTIPLE-OUTPUT CHANNEL MODELS

Wiley Series in Adaptive and Learning Systems for Signal Processing, Communication and Control

Editor: Simon Haykin

Adali and Haykin / ADAPTIVE SIGNAL PROCESSING: Next Generation Solutions
Beckerman / *Adaptive Cooperative Systems*
Candy / *Model-Based Signal Processing*
Chen, Haykin, Eggermont, and Becker / *Correlative Learning: A Basis for Brain and Adaptive Systems*
Chen and Gu / *Control-Oriented System Identification: An H∞ Approach*
Cherkassky and Mulier / *Learning from Data: Concepts, Theory, and Methods*
Costa and Haykin / *Multiple-Input, Multiple-Output Channel Models: Theory and Practice*
Diamantaras and Kung / *Principal Component Neural Networks: Theory and Applications*
Haykin / *Unsupervised Adaptive Filtering: Blind Source Separation*
Haykin * *Unsupervised Adaptive Filtering: Blind Deconvolution*
Haykin and Liu * *Handbook on Array Processing and Sensor Networks*
Haykin and Puthussarypady * *Chaotic Dynamics of Sea Clutter*
Haykin and Widrow * *Least Mean-Square Adaptive Filters*
Hrycej * *Neurocontrol: Towards an Industrial Control Methodology*
Hyvärinen, Karhunen, and Oja * *Independent Component Analysis*
Kristić, Kanellakopoulos, and Kokotović * *Nonlinear and Adaptive Control Design*
Mann * *Intelligent Image Processing*
Nikias and Shao * *Signal Processing with Alpha-Stable Distributions and Applications*
Passino and Burgess * *Stability Analysis of Discrete Event Systems*
Sánchez-Peña and Sznaier * *Robust Systems Theory and Applications*
Sandberg, Lo, Fancourt, Principe, Katagairi, and Haykin * *Nonlinear Dynamical Systems: Feedforward Neural Network Perspectives*
Sellathurai and Haykin * *Space-Time Layered Information Processing for Wireless Communications*
Spooner, Maggiore, Ordóñez, and Passino * *Stable Adaptive Control and Estimation for Nonlinear Systems: Neural and Fuzzy Approximator Techniques*
Tao * *Adaptive Control Design and Analysis*
Tao and Kokotović * *Adaptive Control of Systems with Actuator and Sensor Nonlinearities*
Tsoukalas and Uhrig * *Fuzzy and Neural Approaches in Engineering*
Van Hulle * *Faithful Representations and Topographic Maps: From Distortion- to Information-Based Self-Organization*
Vapnik * *Statistical Learning Theory*
Werbos * *The Roots of Backpropagation: From Ordered Derivatives to Neural Networks and Political Forecasting*
Yee and Haykin * *Regularized Radial Bias Function Networks: Theory and Applications*

MULTIPLE-INPUT, MULTIPLE-OUTPUT CHANNEL MODELS
Theory and Practice

By

Nelson Costa
Simon Haykin

Wiley Series in Adaptive and Learning Systems for
Signal Processing, Communication, and Control

The Institute of Electrical and Electronics Engineers, Inc., New York

A JOHN WILEY & SONS, INC., PUBLICATION

Published by John Wiley & Sons, Inc., Hoboken, New Jersey
Published simultaneously in Canada

For general information on our other products and services or for technical support, please contact our Customer Care Department within the United States at (800) 762-2974, outside the United States at (317) 572-3993 or fax (317) 572-4002.

Wiley also publishes its books in variety of electronic formats. Some content that appears in print may not be available in electronic format. For more information about Wiley products, visit our web site at www.wiley.com.

Library of Congress Cataloging-in-Publication Data:

Costa, Nelson, 1975-
  Multiple-input multiple-output channel models : theory and practice/Nelson Costa, Simon Haykin.
    p. cm.
  Includes bibliographical references and index.
  ISBN 978-0-470-39983-5 (cloth)
  1. MIMO systems. I. Haykin, Simon S., 1931-II. Title
    TK5103.2.C675 2010
    621.382--dc22
                          2009052165

Printed in the United States of America

10 9 8 7 6 5 4 3 2 1

To my parents and my wife,
for their patience and encouragement.

—Nelson Costa

CONTENTS

PREFACE

Many wireless communications channels consist of multiple signal paths from the transmitter to receiver. This multiplicity of paths leads to a phenomenon known as *multipath fading*. The multiple paths are caused by the presence of objects in the physical environment that, through the mechanisms of propagation, alter the path of radiated energy. These objects are referred to as *scatterers*. In the past, researchers often looked at ways to mitigate multipath scattering, such as in diversity systems. *Multiple-input, multiple-output* (MIMO) systems, on the other hand, use multipath diversity to their advantage; a MIMO system has the ability to translate increased spatial diversity into increased channel capacity. This promise of increased capacity has meant that, over the course of the past 10 years, MIMO technologies have become more widespread. MIMO has been accepted into numerous wireless standards, such as the IEEE 802.11n.

The capacity of a MIMO channel is highly dependent on the spatial structure of the channel. MIMO *channel models* are an important tool in understanding the potential gains of a MIMO system. This text presents the theory behind MIMO channel modeling in the context of linear system theory and probability. It discusses examples of two types of MIMO channel models in detail: *correlative channel models* and *cluster models*.

Channel models are validated using data measured from real-life channels. In this way, channel modeling and *channel sounding* are closely related. The text discusses the theory behind different channel sounding techniques, including those used to measure the wideband MIMO channel. MIMO channel sounders can be roughly divided into two categories: *true MIMO* and *switched-array* sounders. The text describes two important examples of wideband MIMO channel sounders; namely, the *wideband MIMO software defined radio* (WMSDR) and the *Brigham Young University* (BYU) wideband channel sounder. The WMSDR is a true MIMO sounder, capable of transmitting and receiving digital data simultaneously on all of its antennas. The BYU sounder is a switched-array type. To address the applicability of several models, real-life data is used to validate their performance. This includes a discussion on metrics used in validating wideband MIMO channel models and their meaning.

Throughout the text, the focus is on a balanced treatment of the theory and application of wideband MIMO channel models. Each chapter includes a list of important references. This includes core literary references, MATLAB implementations of key models, and the location of databases that can be used to help in the development of new models or communication algorithms. The text is intended primarily for engineers at a graduate or postgraduate level. It is intended to give the reader a clear understanding of the underlying propagation mechanisms in the wideband MIMO channel. This knowledge is fundamental to the development of communication algorithms, signaling strategies, and transceiver design for MIMO systems.

ACKNOWLEDGMENTS

It took four years, and more than a little patience, to build the WMSDR from scratch. In particular, the authors wish to thank Dr. Jamal Deen, and his primary researcher Dr. Ogi Marinov (McMaster University), for lending equipment and expertise essential to the building of the WMSDR. Special thanks also go to Dr. Peter Smith (McMaster University) for access to his test equipment. The WMSDR would not have been possible without their support.

Thanks go to Dr. Tim Davidson (McMaster University), who provided much needed feedback on preliminary drafts of the text.

Dr. Tricia Willink (Communications Research Center) provided us with the original Weichselberger model references and background material. This material proved to be invaluable in understanding the ideas behind correlative channel modeling and in developing the structured model. Dr. Mathini Sellathurai (Queens University, Belfast) also provided us with many valuable references and feedback on early versions of the transcript.

The authors are deeply indebted to Dr. Mike Jensen (Brigham Young University), who allowed us access to an extensive wideband channel estimation database at Brigham Young University. The data proved invaluable in the initial development and validation of the structured model.

The authors owe many thanks to Debby Costa and David Lindo for assistance during the WMSDR measurement campaigns. Without their help, we never would have been able to collect the WMSDR data that forms a cornerstone of this text.

A special thanks goes to Dr. Nicolai Czink (Stanford University) and to the people at Electrobit and at Forschungszentrum Telekommunikation Wien (FTW). Dr. Czink provided critical feedback on Chapter 4, as well as the experimental data in Fig. 6.1. The people at Electrobit and FTW funded his research and also agreed to make the MATLAB implementation of the random cluster model (RCM) publicly available.

Finally, thanks are also owed to the National Sciences and Research Council of Canada (NSERC) for their generous funding, for more than six years, of the research that is a cornerstone of this text.

1

INTRODUCTION

We begin with a brief historical account of some of the major contributors to wireless communications. We then list background material on *multiple-input, multiple-output* (MIMO) communications, MIMO channel models, and *software defined radio* (SDR) and include some important references. The chapter concludes with a brief summary of the remainder of the text.

1.1 HISTORICAL PERSPECTIVE

In retrospect, wireless communications in its current form is the result of many discoveries made by many people over hundreds of years. In the literature, a handful of people is credited with the most important discoveries in wireless communication theory. The origins of wireless communication can be traced back to the first experiments on electricity and magnetism.

1.1.1 Electromagnetism

The eighteenth century saw the discovery of electricity, including Benjamin Franklin's famous kite-flying experiments, and the invention of the battery by Alessandre Volta in 1800. In the latter half of the eighteenth century, some people began to suspect that there might be a connection between electricity and magnetism. In 1820, the Danish physicist Hans Christan Øersted first made the observation that he could force a

Multiple-Input, Multiple-Output Channel Models: Theory and Practice. By Nelson Costa and Simon Haykin
Copyright © 2010 John Wiley & Sons, Inc.

compass needle to deflect at right angles to a current-carrying wire. This was perhaps the first concrete evidence that electricity and magnetism were somehow linked (1). Later that same year, a week after hearing about Ørsted's discovery, the French physicist André-Marie Ampère presented a detailed paper on the phenomenon.

From that point forward, physicists such as Michael Farady, André Ampère, and Joseph Henry tried to define the mathematical relationship between electricity and magnetism. Defining this relationship proved to be a difficult task. It was the Scottish mathematician and physicist James Clerk Maxwell who is largely credited with discovering the laws of electromagnetism. Maxwell presented his work over the course of three publications.

In the 1861 publication "On Physical Lines of Force" (2), Maxwell first presented what would eventually be known as *Maxwell's equations*. Arguably, the most important contributions of the paper were the following four laws of electromagnetism:

1. *Gauss's law*, defining the relationship between an electric field and free electric charge;
2. *Gauss's law of magnetism*, which states the net flow of a magnetic field at any point is zero, implying the nonexistence of magnetic monopoles;
3. *Faraday's law of induction*, showing how a changing magnetic field can induce an electric field; and
4. *Ampère's circuital law*, showing how a changing electric field can produce a magnetic field. Except for Ampère's circuital law, the first three laws had already been stated elsewhere. In Ref. 2, using Faraday's *lines of force* and his own model, Maxwell rederived the above relations (3).

In 1864, Maxwell published "A Dynamical Theory of the Electromagnetic Field" (4). In this paper, he expanded on Ref. 2, and presented a set of eight equations, which would become known as Maxwell's equations. In addition to this, Maxwell also correctly predicted that light was an *electromagnetic wave* and followed the same laws as electromagnetic fields. He then presented a wave equation that described the propagation of electromagnetic (EM) waves through a medium. Maxwell implied that, using a time-varying current, it was possible to generate electromagnetic waves that traveled at the speed of light.

In 1873, Maxwell brought all his previous work together in a single book entitled *A Treatise on Electricity and Magnetism* (5). This book is most often cited as Maxwell's most important contribution. The book contained a series of 20 equations that defined the relationship between electric fields, magnetic fields, electric charge, and electric current (3). The book also provided a comprehensive review of Maxwell's findings in EM theory.

Maxwell's discoveries were so advanced that it took more than a decade for people to realize their importance. In 1884, the English electrical engineer Oliver Heaviside, after reviewing Maxwell's work, selected four of Maxwell's original equations and rewrote them using modern differential notation. The revised equations are sometimes referred to as *Heaviside equations* to differentiate them from the original eight equations given in Ref. 4. Heaviside's simplification brought Maxwell's discoveries to the forefront and led to the development of the first wireless transmitters and receivers.

1.1.2 The Hertz Transmitter

The person most often credited with building the world's first wireless transmitter was the German physicist Heinrich Hertz in the late nineteenth century. From 1885 to 1889, Hertz was a professor of physics at Karlsruhe Polytechnic. During this time, Hertz was trying to find proof of EM waves, as predicted by Maxwell. In 1886, Hertz developed the *Hertz antenna receiver*, which is illustrated in Fig. 1.1. Hertz built his transmitter using a Ruhmkorff-type induction coil connected to a center-fed dipole equipped with a spark gap. Capacitive elements at the transmit antenna helped to tune the frequency of the transmitted energy. The induction coil caused sparks to appear at the transmit antenna gap. The receiver consisted of a simple loop antenna, equipped with a similar spark gap. During operation, the transmitter would cause sparks to appear at the receive-antenna gap when the transmitter was separated from the receiver by a small distance. Over the course of numerous experiments, Hertz was able to measure the wavelength and velocity of the EM waves. He demonstrated that the EM waves traveled at the speed of light, despite having a much larger wavelength than light. Hertz also experimented with EM propagation through different media and showed that the EM energy would transmit through some media and would reflect off others. He demonstrated that the *refraction* and *reflection* properties of EM waves were similar to those of light, as predicted by Maxwell. This discovery also formed the bases for *radar* technology, among other things (6).

Not even Hertz realized the practical implications of his invention. He is quoted as saying that his apparatus was "of no use whatsoever [. . .] this is just an experiment that proves Maestro Maxwell was right—we just have these mysterious electromagnetic waves that we cannot see with the naked eye" (6). Despite his apparent lack of vision, Hertz's discoveries inspired an entirely new field of research. He demonstrated that EM waves could be used to induce an action-at-a-distance but failed to see the

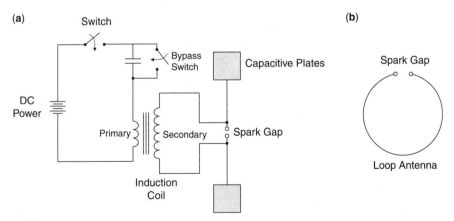

Figure 1.1 Hertz's spark-gap experimental setup showing (a) transmitter and (b) receiver.

significance of this discovery. Hertz's discoveries led directly to the development of the first wireless telegraphing devices.

1.1.3 Tesla and Wireless Power

It took another decade for people to make the connection between Hertz's work and wireless communications. Motivated primarily by his research into wireless power transmission, in 1891 Nikola Tesla demonstrated the wireless propagation of power to vacuum tubes. He constructed several different alternators that generated AC power between 15 and 18 kHz. He used the alternators to wirelessly power phosphorus-lined vacuum tubes and cause them to glow. In 1892, at a meeting of the Institution of Electrical Engineers of London, Tesla stated that this technology could be used to convey information from one point to another wirelessly (7). As early as 1896, Tesla transmitted a continuous-wave signal from his Houston St. lab in New York City to West Point, a distance of approximately 48 km. The transmitter consisted of an alternator, which generated a continuous wave at approximately 5 kHz. The receiver was made up of a steel wire, two condensers, and a strong electromagnet. The receiver was tuned to the transmitter by placing the wire within the magnetic field, thus forming a resonant circuit in conjunction with the condensers (7). In 1943, soon after Tesla's death, the Supreme Court of the United States overturned one of Marconi's patents and named Tesla "the inventor of the radio" (8).

1.1.4 Lodge and Tunable Circuits

In 1890, French physicist Édouard Branly showed that metal filings in a glass tube would "cohere," or clump together, when under the influence of EM waves. The English physicist Oliver Lodge improved on Branly's radio detector by including a "trembler" that would dislodge the filings between cycles, thus maintaining the receiver's sensitivity. He named the device a *cohere*. Lodge connected the cohere to a receiver circuit that included an inker to record the received signals. In 1894, at a meeting of the British Association for the Advancement of Science at Oxford University, Lodge used his apparatus to transmit and receive Morse-encoded signals across a distance of 150 m, thus demonstrating its usefulness in wireless communications.

Perhaps Lodge's greatest contribution was in the area of tunable resonant circuits for communications. While at University College, Liverpool, Lodge experimented with tunable circuits using inductors and capacitors. He showed that untuned circuits would respond to EM waves at almost any frequency, but that their response would fall off quickly. Conversely, a tuned circuit would respond to waves whose frequency corresponded with the circuit's natural frequency. Furthermore, the response would not decay as quickly as for an untuned circuit. In 1898, Lodge submitted a patent for a tunable transmitter and receiver set. The transmitter antenna was equipped with an adjustable induction coil that could be tuned to the receiver. The receiver used his version of Branly's cohere. His "syntonic" patent showed the importance of tuning the transmitter to the receiver. This patent was later acquired by the Marconi company in 1912 (9).

1.1.5 Marconi and Trans-Atlantic Communication

At the beginning of the twentieth century, the wired telegraph was already a widely accepted form of communication. Trans-Atlantic cable had already been laid, making telegraph communication possible between North America and Europe. The disadvantage of cable, however, was that it was difficult and expensive to lay and maintain. Also, the fact that it required both users to be linked by wire precluded its use on ships. Guglielmo Marconi is one of the first people credited with proving the practicality of wireless communication. Marconi took ideas from visionaries before him, such as Hertz and Tesla, Branly and Lodge (among others), and refined them to increase the distance between transmitter and receiver. He was interested in developing a telegraphing apparatus for commercial and military use.

Marconi started doing experiments in his parents' attic in Pontecchio, Italy, at a young age. By 1895, he demonstrated the wireless transmission of a Morse-encoded message across a distance of 1.5 km. At this point, Marconi concluded that he needed more funding to extend the range of his apparatus. In a bid to solicit financial support, Marconi traveled to London, England, in 1896. It was there that Marconi first demonstrated his apparatus to the British Post Office, thereby gaining the attention of the British government. From 1896 to 1899, the British government funded a series of wireless experiments. With each new experiment, Marconi refined his apparatus and increased the distance between the transmitter and receiver. In 1897, Marconi was the first to transmit a signal successfully over a body of water. The Morse-encoded signal crossed the Bristol Channel, from Lavernrock Point in South Wales to Flat Holm Island, a distance of approximately 14 km. By 1899, Marconi had successfully broadcast signals across the English Channel, from Wimereux, France, to the South Foreland Lighthouse in England, a distance of more than 45 km (10, 11).

In 1901, Marconi made his famous trans-Atlantic transmission, in which the Morse-encoded letter "S" was broadcast from Poldhu, England, to St. Johns, Newfoundland, Canada, across a distance of approximately 3500 km. At the time, some questioned the validity of Marconi's experiment. Because Marconi's device used a cohere detector at the receiver, the transmitter used a low spark rate (estimated later to be of the order $2-3$ sparks/s); this ensured that the detector could reset itself between cycles. At the receiver, Marconi was barely able to hear the transmitted signal above the background atmospheric noise. Critics claimed that the low spark rate, combined with the very low signal-to-noise ratio, made it impossible to distinguish the transmitted signal from background noise. They concluded that Marconi had, in fact, only heard the latter. In a bid to silence critics, Marconi set up a mobile experiment to measure the maximum distance of the receiver. In 1902, he set up his cohere receiver on the S.S. *Philadelphia* and sailed west from the Poldhu station, recording the received signal as he went. He found that the maximum range of the transmitted signal was 2496 km at night and only 1125 km during the day, less than half the distance from Poldhu to St. Johns. The St. Johns–Poldhu experiments had been conducted during the day. Although this experiment cast an even bigger shadow over his 1901 claim, Marconi did prove that signals could propagate hundreds of kilometers. Until then, the prevailing view was that signals could only travel within line-of-sight (10, 11).

Despite doubts about his most famous claim, Marconi was very successful at convincing the public of the importance of wireless communications. He leveraged his influence with the British and American governments to build high-power transmission stations on both sides of the Atlantic and to build at least two successful companies. In 1904, he established a news telegraph service for seafaring ships, and 1907 was the first year that his company provided regular telegraph service across the Atlantic. In 1909, in conjunction with Karl Ferdinand Braun, Marconi won the Nobel Prize for Physics "in recognition of their contributions to the development of wireless telegraphy" (11).

1.2 MIMO COMMUNICATIONS

The field of wireless communications has grown since the days of Marconi's first wireless transmission across the Atlantic. Wireless devices are now ubiquitous in our society. From cell phones to satellite links, they have become a necessary part of everyday life. Wireless technology allows us to communicate across large distances, such as in satellite communication. It also allows the wireless devices themselves to communicate with each other, such as a Bluetooth handset with a cordless ear piece. With the proliferation of wireless devices, the goal has shifted from making wireless communication work to making it more efficient. Given the proliferation of communication devices and a limited set of network resources, we wish to increase the number of *users* sharing the network while increasing the reliability of each user's link to it.

Perhaps the most precious resource in a given radio channel is the amount of available *bandwidth*, or the frequency range that it occupies. Regardless of the communication scheme, the number of users one can place within a given bandwidth is limited. Government bodies strictly regulate the allotment of bandwidth to civilian entities, such as cell phone service providers. Given a constant transmission rate for each user in a network, providers want to maximize the number of users within their allotted bandwidth. Thus, the issue then becomes one of *bandwidth efficiency*.

Since the time of Marconi, it was found that, typically, the signal strength at a receiver varies greatly with small changes in its position. The prevailing view is that the received signal is composed of many signals arriving from many different directions. Each signal is formed when the transmitted energy takes a different *path* from the transmitter to the receiver. Through propagation mechanisms such as reflection, diffraction, and scattering, objects in the channel create multiple paths from the transmitter to the receiver. We collectively refer to these objects as *scatterers*. The paths are of different lengths, and thus the signals arrive at the receiver with different amplitudes and phases. This is illustrated in Fig. 1.2. In some cases, the multiple signals add destructively at the receiver, creating points in space where the composite received signal is greatly attenuated. This is referred to as *multipath interference*. To combat the effects of multipath interference, we can employ an array of antennas at the receiver, with each antenna separated by some distance in space. This is illustrated in Fig. 1.3 for the case where the receiver employs a two-element *antenna array*.

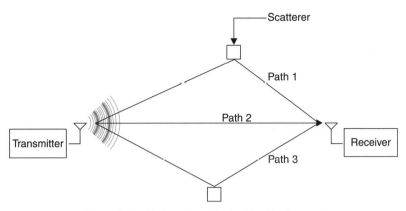

Figure 1.2 Abstraction of a multipath channel.

In this way, we increase our chances that at least one antenna at the receiver does not suffer from multipath fading. By increasing the number of antennas at the receiver, we better the odds. Also, if we combine the signal from both antennas intelligently, we can increase the overall signal strength at the receiver. This is an example of *receive-diversity*. We can also increase the number of antennas at the transmitter to combat multipath fading. This is an example of *transmit-diversity*. In general, diversity techniques greatly improve the average received signal strength in multipath channels. Diversity systems are a fairly mature topic in wireless communications. Jakes (12) provides one of the best introductions to multipath fading and diversity systems. A more recent review of diversity system techniques is given in Refs. 13 and 14.

Work in diversity systems naturally led to the inception of MIMO systems. In channels with rich multipath, the probability that we have more than one independent path increases. This independence is the key to diversity systems; if one path exhibits a deep fade, rich multipath ensures that the likelihood of the other paths doing so at the same

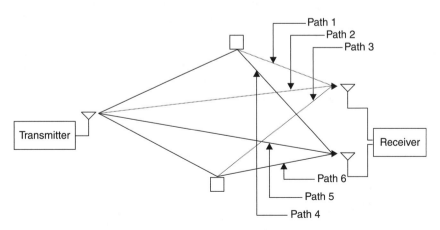

Figure 1.3 Example of a 2 × 1 receive-diversity system.

instant is small. Having the ability to resolve independent paths in the channel also implies that we can increase the amount of information we transmit by sending multiple signals into the channel *within the same bandwidth.* In this way, MIMO systems add another degree of freedom; whereas diversity systems use multipath to increase the *reliability* of a link, MIMO systems can trade reliability for an increase in link *capacity.* Consider the case shown in Fig. 1.4, where the transmitter and receiver are both equipped with a two-element array.

At the transmitter, each antenna broadcasts an independent signal on the same bandwidth. The two signals are combined in the channel. Each receive-antenna captures an independently faded version of the combined signals. Until about 20 years ago, because the signals were combined in the channel, it was impractical to separate them at the receiver. It was Winters who first suggested that, by employing multiple antennas both at the transmitter and receiver, we could increase the number of transmitted signals to increase the capacity of a given channel (15). He did not, however, indicate a coding method by which we could separate the transmitted streams at the receiver.

It was not until 1996 that Foschini introduced a coding algorithm that took advantage of the added capacity in a MIMO channel (16). The algorithm encoded data across time and across all transmit antennas. The result was one of the first *space–time algorithms.* The algorithm was later dubbed *Bell LAbs layered Space–Time* (BLAST). It was also in this paper that Foschini first presented the now-famous capacity formula for MIMO channels,

$$C = \log_2 \det\left[\mathbf{I}_{M_{Rx}} + \frac{\rho}{M_{Tx}} \mathbf{HH}^H \right] \text{b/s/Hz}, \tag{1.1}$$

where $\mathbf{I}_{M_{Rx}}$ is the $M_{Rx} \times M_{Rx}$ identity matrix, ρ is the *system signal-to-noise ratio* (SNR), $(\bullet)^H$ is the Hermitian transpose, M_{Rx} and M_{Tx} denotes the number of receivers and transmitters, respectively, and $\mathbf{H}$ is a matrix whose entries are the $M_{Rx} \times M_{Tx}$ channel gains, often referred to as the *H-matrix.* In Ref. 17, Telatar gives a detailed analysis of the capacity in multiantenna Gaussian channels. Perhaps the simplest space–time algorithms is the Alamouti space–time code (18), which is applicable to the 2×2 case shown above. In the Alamouti scheme, information is coded across both antennas, and over two symbol periods, thus across both time and space.

Foschini showed that, under ideal conditions, the data rate of his algorithm approached the capacity limit of the MIMO channel. These limits were investigated more thoroughly in Ref. 19. Foschini's BLAST touched off an entirely new field of

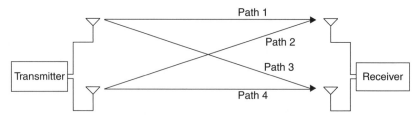

Figure 1.4 Example of a 2 × 2 MIMO system.

study in space–time coding. Since then, many other space–time algorithms, such as vertical BLAST (V-BLAST) (20) and Turbo-BLAST (21), have been developed. MIMO theory has also had far-reaching consequences in other areas of communications, such as multiuser wireless networks (22) and even optical fiber communications (23). MIMO is arguably one of the most important single advances in communications in the past decade.

Around the time of Foschini's landmark paper, MIMO was largely touted as a method by which we could increase the capacity of a given link without increasing the bandwidth; that is, we increase the *spectral efficiency* of the channel. As discussed above, multiple antennas at the transmitter or receiver can also be used to increase the *reliability* of a wireless link via diversity. More recently, it has been shown that we can trade capacity for diversity to increase the reliability of the wireless link. This trade-off represents an added degree of flexibility and is known as the *diversity-multiplexing trade-off* (24, 25).

1.3 MIMO CHANNEL MODELS

In the past, the increased capacity of a MIMO versus *single-input, single-output* (SISO) channel was often shown using the assumption that individual paths in the MIMO channel are *independent*. Over the course of many experimental studies, it was found that, in many practical cases, this assumption was not accurate. Experimental data have shown that, depending on the environment, the capacity of a measured channel often falls short of the limit given by Foschini (26–28). Space–time algorithms that focus on approaching the MIMO capacity bound rely heavily on the multipath diversity of the channel. As a result, their overall performance is highly dependent on the environment. Thus, there is need to accurately model the MIMO channel to evaluate properly the performance gains of MIMO versus diversity or single antenna systems.

With respect to MIMO system development, MIMO channel models serve a twofold purpose. First, a MIMO channel model can be used in the design of a MIMO system. This includes the design of an optimal signaling scheme, detection scheme, and space–time code. The same model can often be used to test a given system. In this case, the model acts as a *channel simulator*. We can generate exemplar channels using the model and use them to test the performance of a given system. The second and perhaps most important purpose of a MIMO channel model is to gain some insight into the underlying physics of the channel. At a higher level, an accurate channel model can tell us a lot about the behavior of a given channel. It imposes a structure that allows for a deeper understanding and can indicate what assumptions can safely be made to optimize system design.

1.3.1 The Channel Model Spectrum

In general, most wireless channel models fall somewhere in the spectrum indicated in Fig. 1.5. Models on the left of the spectrum are closely related to the channel physics. They rely on EM propagation techniques to solve for EM field values given specific

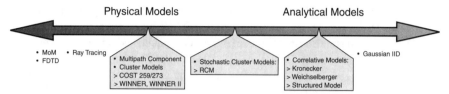

Figure 1.5 The spectrum of channel models, indicating where a few of the MIMO channel models discussed in this text fall.

geometries. These models are often the most accurate but are by far the most complex. *Finite difference time domain* (FDTD) (29, 30) models solve for Maxwell's equations given specific boundary conditions and are most often used on small scales. *Ray tracing models* are based on EM propagation mechanisms such as free-space propagation, reflection, diffraction, and scattering (31). They require a complete description of every object in the channel, including doors, walls, desks, and so forth. This includes each object's geometry, location, and scattering properties.

Cluster models are a compromise between stochastic and ray-tracing models. They use rays, called *multipath components*, to simulate reflected energy in a channel. The multipath components are stochastically modeled using a number of parameters measured from real-life channels. To simplify the model, the multipath components are grouped into *clusters*, with multipath components in each cluster having similar statistics. Clusters, in turn, simulate the behavior of objects in the channel.

Scattering models use the concept of virtual scatterers and simplified ray tracing to capture the statistical properties of the channel (32–34). Many of these models begin with a simple EM ray-based propagation model, such as that described by Jakes (12), choose a scatter geometry that approximates the behavior of scatterers in a real environment, and then derive relationships for the correlation or capacity as a function of physical parameters. Prior geometries include a ring of scatterers around the receiver only (31), one line of scatterers at the transmitter and receiver (32), and layers of scatterers (33). Emphasis is usually placed on correlation between antennas and the resulting capacity distribution of the channel given some antenna spacing.

Further to the right-end of the spectrum, *correlative models* attempt to approximate the *spatial structure* of the channel by modeling the correlation between paths in a MIMO channel. These models bias a "white" channel matrix with correlation matrices at the transmitter and receiver to approximate the correlation between scatterers at both link ends. The white channel matrix consists of uncorrelated complex-Gaussian elements (35–37).

In general, channel models for wireless systems have been primarily concerned with modeling the effects of the following parameters on the transmitted signal: path loss, delay spread, shadowing, Doppler spread, and Ricean K-factor. Parameters of specific interest to the MIMO channel are the joint antenna correlations and channel matrix singular-value distribution. Also of interest in the MIMO channel is the spatial diversity, which can be measured by analyzing the H-matrix singular-value distribution (38). In this way, the MIMO channel can be decomposed into a number of parallel SISO channels whose channel gains are directly related to the singular values of the H-matrix. By analyzing the distribution of the singular values in different environments, we can assess the performance gain of MIMO over other signaling techniques.

1.3.2 Wideband MIMO Channel Models

Increasing the bandwidth of a wireless channel often means that it begins to affect a given signal differently at different frequencies. When *fading* is *frequency dependent*, we refer to the channel as being *wideband*. Wideband channel characterization, in the case of single antenna systems, is a fairly mature topic. The first characterizations of the wideband channel appeared in the works of Kailath (39) and Bello (40). Both describe the wideband channel as a linear time-varying filter and apply linear theory to its characterization. *Bello's model* characterizes both the time and frequency variation of a wideband channel. Since then, there have been many experiments to validate his model and measure the parameters of different real-life channels. See, for example, Refs. 41–44, among others.

In contrast, the characterization of wideband MIMO channels is a relatively new field of study. To date, only a handful of wideband MIMO test beds exist. Some examples are given in Refs. 45–49. There are even fewer wideband MIMO channel models. Due to the complexity and difficulty of building a wideband MIMO channel sounder, fewer still test their channel model with real-life data. One of the most notable works in the field of wideband MIMO channel modeling is that of Yu et al., in which a known narrowband correlative model, called the *Kronecker model*, is extended to the wideband case (36). Yu et al. show the Kronecker model to be fairly accurate in predicting the correlation between paths in a MIMO channel. However, more recent work suggested weaknesses in the Kronecker model, especially as the number of antennas at the transmitter and receiver is increased (50, 51).

To reduce the number of model parameters, the Kronecker model assumes that scatterers at the transmitter are not correlated with those at the receiver. We refer to this as the *separability assumption*, as, in effect, it implies that the effects of scattering around the transmitter can be separated from those around the receiver. In Ref. 50, Özcelik et al. argue that this assumption is a source of modeling error. They also imply that, for many real-life channels, the scatterers at either link end are indeed coupled in some way.

In Chapter 3, we present a relatively new wideband MIMO channel model, called the *structured model*. The model is unique for the following reasons. It models the coupling between scatterers at either link end. It models the correlation between scatterers at different delays. The model also recasts the wideband MIMO channel gains as a third-order tensor. The model was derived using *tensor decomposition*. In Chapter 6, using real-life data, we compare the performance of the structured model versus the Kronecker model. We show that the structured model better approximates the spatial structure of a given channel versus the Kronecker model.

1.4 SOFTWARE DEFINED RADIO

Over the past decade, the number of wireless communication standards has grown. Currently, there are many competing cellular telephone standards, such as *Code Division Multiple-Access* (CDMA) and *Global System Mobile* (GSM); wireless computer networks such as the IEEE 802.11n and Bluetooth; as well as a host of military standards. Different standards have different hardware requirements; the carrier frequency and the method by which the signal is encoded will differ across standards, requiring different antennae, filters, mixers, and so forth. Often, a single device will be designed to work across a variety of network standards. For example, many cell phones are designed to work in both North American and European standards. In many cases, it would be advantageous for a single device to have the ability to communicate using multiple standards. Traditionally, doing this meant that we had to increase the hardware complexity of the device, thereby increasing its development time and unit cost.

Most communications devices nowadays perform their task with a combination of hardware and software. The hardware translates analog signals in the channel to digital signals that can be processed in software. Its functionality is fixed; once implemented, it is difficult to alter its specifications. By contrast, the software functionality is often more malleable and can be changed by simply writing more software. Thus, the goal of SDR is to reduce the amount of hardware by increasing software functionality, thereby increasing overall flexibility.

The dichotomy between hardware and software is best illustrated by way of example. Figure 1.6 shows a block diagram of a typical digital receiver. Radio frequency (RF) signals are captured by an antenna and fed to a *RF conversion* block. The output of this block is connected to an *analog-to-digital converter* (ADC), which converts the analog signals to digital form. Once in the digital domain, the signals can be processed in many ways using software. In practice, the input frequency and amplitude range of ADCs are limited. The RF signals, on the other hand, are usually centered at a relatively large frequency, which we refer to as the *carrier frequency*. Furthermore, there are many other signals adjacent to the signal of interest in the frequency domain. Therefore, the RF conversion block consists of all *hardware* responsible for converting the frequency and amplitude range of the RF signal of

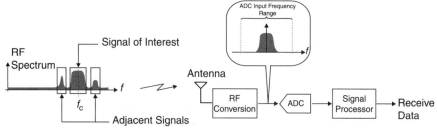

Figure 1.6 A typical digital receiver, highlighting the division between the analog and digital domains.

interest to a range that is compatible with the ADC and filtering out the undesired signals. The signal processor block is responsible for decoding the signal and estimating the transmitted message. Depending on the capability of the hardware before this block, including the ADC, and depending on the capabilities of the signal processor itself, we can decode just about any type of signal using software.

In the above example, the input range of the ADC determines, to some extent, the specifications of the RF conversion-block. That is to say, if the ADC input range is small, we must increase the complexity of the RF conversion block to decode a given signal. In Ref. 52, Tuttlebee distinguishes between *pure* and *pragmatic* SDR. In its purist form, a software defined receiver would consist of an antenna, ADC, and signal processor block, as shown in Fig. 1.7. In this case, the input range of the ADC is such that it would capture the entire frequency spectrum, including the signal of interest, and convert everything to the digital domain. That is to say, the input amplitude and frequency range would be large enough to capture all signals in the entire RF spectrum. In this way, we can write software to isolate and decode the signal of interest, regardless of where it lies in the RF spectrum.

In most cases, implementing a pure SDR is impractical, if not impossible. Even with the ever-increasing capabilities of data conversion technology, the RF spectrum contains many undesired signals. Government regulations force providers, such as cell phone and wireless data providers, to occupy relatively small, well-defined segments of spectrum. Also, converting the entire spectrum to the digital domain requires enormous computing power. For example, consider the case where we have a 32 kHz signal, centered at 5 GHz. To capture this signal, an ADC would have to operate at a sample rate greater than 10 GHz. Assuming 8-bits/sample, a computer would have to process greater than 10 MB/s just to capture a 32 kHz signal. This data rate increases linearly as we increase the number of antennas; for example, in MIMO systems. Furthermore, the received signal power can vary widely, especially in mobile applications. The dynamic range of the ADC, determined by the number of bits at its output, is often much smaller than the range of the received signal. The received signal is lost while its power falls outside of the ADC input range.

For these reasons, pragmatic SDR tries to strike a balance between pure hardware radios with limited functionality and pure SDRs, which are impractical. As technology advances, more and more functionality is moved to the signal processor. Practical considerations such as portability, battery life, and cost determine exactly where the line is drawn between hardware and software functionality.

Figure 1.7 The purist SDR.

Mitola (53) is the person most often credited as having coined the term "software radio." In Ref. 54, he gives a detailed vision of SDR in the context of a cellular network. Mitola describes all the necessary functions that a software radio would have to perform to communicate with the network. His vision is that of a pragmatic SDR; the device would consist of RF conversion hardware and a signal processor. Every stage of the RF conversion is controlled by the signal processor in real time. The requirements of each function block, from the RF signal to the decoded message, are described in detail. The requirements are mostly derived from cellular standards such as CDMA and GSM.

One of the more intriguing contributions of the paper involve the idea of quantifying the *total available resources*, like computing power, and dividing these resources quantitatively between different functions, depending on the task. For example, an analog voice signal requires different resource allocation versus a CDMA signal. These resources are dynamically allocated by the signal processor in real time. Mitola states that this is one of the distinguishing features of a software radio versus a "software-controlled" radio, although he does not quantify the distinction.

Since Mitola's paper was first published, the definition of SDR has undergone many changes. The ideas contained in Ref. 53 have also led to the creation of the *SDR Forum*, whose body is "dedicated to promoting the development, deployment and use of software defined radio technologies for advanced wireless systems" (55). The forum is composed of individuals from many different companies, universities, and government bodies all over the globe. Specifications for software radio development have already been included in some standards, such as the *3rd Generation Partnership Project* (3GPP) (56).

Although the definition has changed, the goal of SDR remains the same; to implement a SDR, we want to move as much functionality as possible to the digital domain and give the signal processor as much control over the remaining RF conversion as possible. In the following, we present an example of a pragmatic SDR, the *Wideband MIMO Software Defined Radio* (WMSDR). The WMSDR was built at McMaster University for the purpose of characterizing outdoor wideband MIMO channels. It consists of a separate transmitter and receiver. The transmitter is equipped with four independent transmit chains. Each chain can be controlled in software. The receiver is similarly equipped with four independent receive chains. We use software to record and process the information from each chain. In this way, the WMSDR is a highly versatile 4×4 SDR, capable of many tasks.

1.5 OVERVIEW

The following section outlines the contents of the remainder of the text.

1.5.1 Chapter 2: Multiple Antenna Channels and Correlation

Chapter 2 describes some of the fundamentals of multiple antenna channels. Most wireless channels can be characterized using linear system theory. This chapter defines the different classes of channels, including the time-invariant narrowband channel, leading through to the time-varying wideband channel. It is here that we introduce the tensor system model for the wideband MIMO channel. The third-order *H-tensor* is an elegant way of describing the complex gains in a wideband MIMO channel and leads to channel tensor decomposition.

Of importance to MIMO channels in general is the concept of *channel correlation*. We define correlation in narrowband and wideband MIMO channels and discuss its *eigenvalue decomposition* (EVD). We extend the definition of correlation to include the wideband MIMO channel using tensor calculus. Using the *azimuth power spectrum* (APS), we define the *spatial structure* of the channel and show how this relates to the channel correlation. Both the channel correlation and its EVD are fundamental to the correlative models covered in the next chapter.

1.5.2 Chapter 3: Correlative Models

Chapter 3 describes the first class of MIMO channel models, correlative models. Specifically, we review the *Kronecker*, *Weichselberger*, and *structured* models. We first outline the fundamentals of correlative modeling, including how correlation can be used to generate an ensemble of new channels with the same spatial structure as a given channel. The concept of *one-sided correlation* is used in all three models to reduce the number of parameters needed to describe the channel. The Kronecker model gets its name from the fact that it approximates the full correlation as the Kronecker product of the one-sided correlation matrices. Both the Weichselberger and structured models use the EVD of the one-sided correlation as parameters. The structured model is the wideband extension of the Weichselberger model. In deriving the structured model, the concept of one-sided correlation is extended to include the wideband channel.

The Kronecker model assumes that fading at the transmitter is not linked to that at the receiver. The Weichselberger model (57) suggests the opposite; scatterers at both link ends are coupled in some way. It characterizes this coupling as the average energy coupled between the eigenvectors of the one-sided correlation matrices. Using tensor algebra, the structured model extends this concept to the wideband MIMO channel. The structured model uses a *coupling tensor*, in addition to the eigenbases of the one-sided correlation matrices as input parameters.

1.5.3 Chapter 4: Cluster Models

Chapter 4 provides some of the fundamentals behind cluster models, another class of MIMO channel models. Cluster models address an important shortcoming of correlative models, namely time-variation. They are also more prominent in wireless standards than correlative models.

The chapter covers several important cluster models, leading from the simplest to more complex. Thse include the Saleh–Valenzuela model, the extended Saleh–Valenzuela model, the *European Cooperation in the field of Scientific and Technical Research* (COST) 273 model, and finally the *random cluster model* (RCM).

The Saleh–Valenzuela model was the first to describe the arrival of energy in the delay domain using clusters. Subsequent cluster models extended the Saleh–Valenzuela model to include the *angle of arrival* (AoA) and *angle of departure* (AoD). In this way, clusters were located in space as well as time. The COST 273 model is complex, consisting of many parts. The chapter focuses the discussion on the model implementation. This allows some insight into the many mechanisms that make up the model and explores the reason behind some of its parameters. The RCM is an extension of the COST 273 model. It greatly reduces implementation complexity by characterizing the channel using a multivariate probability density function (PDF). This PDF includes a characterization of time-variation in the channel.

1.5.4   Chapter 5: Channel Sounding

To estimate the impulse response of real-life channels, we employ a technique called *channel sounding*. This chapter presents the theory behind a few channel sounding techniques used to sound narrowband and wideband channels. One of these, *correlative channel sounding*, is used to measure the wideband channel. A discussion of the role of *maximal-length sequences* (ML sequences) in correlative channel sounding is provided. The technique of *digital matched filtering* and how it can be used to estimate the impulse response of a wideband channel is discussed. We then extend digital matched filtering to include the MIMO case. Using real-life data, we show that MIMO matched filtering provides very accurate estimates of the channel gains.

Because of the proliferation of cheap RF signal generators, sampled spectrum sounding is a popular technique for sounding the wideband channel. Switched array sounders are also popular, because they reduce the bandwidth requirement for MIMO channel sounders. We list advantages and disadvantages of both.

To sound the channel, we employ the use of a *channel sounder*. The chapter discusses the WMSDR, an example of a 4×4 wideband MIMO channel sounder. In any digital communication system, including PN-sequence sounders such as the WMSDR, there are several steps required at the receiver to recover the transmitted signal. This involves *timing* and *carrier recovery*. The chapter discusses a few existing timing and carrier recovery algorithms that were implemented in the WMSDR. Examples of their performance are provided.

1.5.5   Chapter 6: Experimental Verifications

Chapter 6 contains a performance analysis of two correlative channel models; namely, the structured and Kronecker models. The chapter begins with a discussion of several important metrics used in the literature to validate MIMO channel models. Using real-life data is an important step in validating any channel model because it shows how useful a given model is in approximating real-life channels. In addition to the WMSDR data, we test both models with data from *Brigham Young University's* (BYU's) 8×8 wideband MIMO channel sounder. We use data from these very different apparatuses gathered in very different environments to highlight some of the shortcomings of the Kronecker model and the robustness of the structured model.

The chapter describes the BYU sounder architecture, specifications, and theory of operation. It highlights the differences between the WMSDR and BYU data sets. The experimental setup for the WMSDR and BYU data sets is described. We compare the performance of the structured model versus the Kronecker model using several important metrics. In the end, we show that the structured model outperforms the Kronecker model when predicting the capacity of a given channel and in approximating the spatial structure of the channel. In all cases considered here, the structured model uses fewer parameters than the Kronecker model. This implies that the channel structure imposed by the structured model is accurate for a wide variety of channels.

We comment on the importance of modeling the correlation between channel gains at different delays by comparing the wideband correlation matrices of real-life channels. We discuss a metric that quantifies the correlation between paths at different delays. This becomes significant, especially as we increase the system bandwidth, and thus should not be ignored when modeling the channel.

1.5.6 Appendixes: Background and Definitions

The text contains additional material, divided into three appendixes. Appendix A provides a short primer on tensor calculus. Appendix B provides proofs used in the derivation of the structured model. This is closely related to material covered in Chapter 3. Finally, Appendix C summarizes the COST 273 channel model, which is discussed in Chapter 4.

2

MULTIPLE ANTENNA CHANNELS AND CORRELATION

In this chapter, we define the terms *wideband* and *MIMO channel*. Here, we summarize several different classes of channels, starting with the narrowband SISO channel and leading to the wideband MIMO channel. We also define *correlation* in a multiantenna channel. We discuss the H-tensor, which is an elegant way of representing the channel gains in a wideband MIMO channel. Characterizing the wideband MIMO channel as a third-order tensor is a small but significant step; we use *tensor calculus* to compute the correlation between paths in the wideband MIMO channel. The correlation is expressed as a sixth-order *wideband correlation tensor*. Expressing the channel correlation using tensor calculus maintains more of the channel structure.

In Appendix A, we provide a brief introduction to tensor calculus. We also introduce notation that is used throughout the rest of the text. Whereas it is not necessary to have a complete understanding of tensor calculus, it is important to be familiar with the notation and operations covered in Appendix A to appreciate the following discussion.

2.1 THE RADIO CHANNEL: DEFINITIONS

2.1.1 The Physical Channel

Communication systems are designed to convey information from a transmitter, across a channel, to a receiver. There are many possible *physical channels*. Some examples

Multiple-Input, Multiple-Output Channel Models: Theory and Practice. By Nelson Costa and Simon Haykin
Copyright © 2010 John Wiley & Sons, Inc.

include an air interface, a coaxial cable, a telephone twisted pair, an optical cable, a waveguide, and so forth. In the following, we are concerned primarily with the *wireless channel*. We assume that the transmitter broadcasts electromagnetic waves that propagate across some air interface to the receiver. In practice, the transmitter radiates energy using an *antenna*. Similarly, the receiver collects energy using another antenna. Both the transmit-antenna and receive-antenna affect the way in which energy is radiated and collected, respectively. The composition of the channel varies widely from one location to the next. Here, we term the *physical channel* to be the continuum of propagation paths from the transmitter to the receiver at any given instant (51). This includes the effects of the transmit-antenna and receive-antenna.

The propagation of EM energy through a medium is governed by Maxwell's equations. The four main propagation mechanisms are *free-space propagation, reflection, diffraction*, and *scattering*. In this way, not all energy radiated by the transmitter makes its way to the receiver. Free-space propagation results when energy from the transmitter makes its way directly to the receiver via a *line-of-sight* path. The three other mechanisms are caused by objects in the channel, such as walls, the ground, cars, trees, and so forth. We collectively term all objects in the channel as *scatterers*.

Each path from the transmitter to the receiver introduces some finite delay. This introduces a phase shift that depends on the propagation length of a given path. In addition, the energy radiated by the transmitter is also subject to additive noise. The energy from all paths adds up at the receiver. The result is that the receiver will often "see" a distorted, noisy version of the transmitted signal. The effect of a linear channel on the transmitted energy is characterized by the channel *impulse response*. The impulse response can take many forms, depending on how we view the channel analytically.

2.1.2 The Analytical Channel

Consider the wireless channel model illustrated in Fig. 2.1. Physically, if we know everything about the channel between the transmitter and receiver antenna ports, including antenna patterns, the location and composition of all scatterers, and so forth, we could use Maxwell's equations to predict the behavior of all EM waves in the physical channel at all frequencies and all time, and thus compute the impulse response. However, given that it is impossible to know all of this information, much less have the ability to solve for all possible boundary conditions in a real-life channel, we have to make some assumptions to make the problem mathematically tractable.

It is important to note that, for a static channel, the physical channel does not change, but our interpretation of it varies widely depending on the signaling strategy we use. Here we make the distinction between *physical* and *analytical* channels.

For example, in the following we discuss whether a channel is *narrowband* or *wideband*, depending on how we consider the effects of time dispersion. A narrowband communication *system* cannot resolve the multiple paths in a physical channel, but a wideband communication *system* can, given the *same* physical channel. Put differently, the narrowband system does not need to consider the time-dispersive nature of

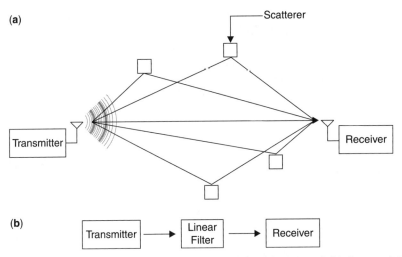

Figure 2.1 Diagram abstraction of (a) the physical channel and (b) the analytical channel.

the channel, but a wideband system does. The physical channel itself does not change, but the way we analyze the same channel in both cases does.

Channel models approximate the behavior of the channel in that they usually focus on only modeling a subset of channel properties. The properties are dictated by the communication system we wish to implement. In the above example, the communication engineer would require a narrowband channel model to design a narrowband communication system effectively and a wideband channel model in the design of a wideband communication system. Thus, in the following, when we speak of *the channel*, we refer to the *analytical channel* and not the *physical channel*.

2.2 CHANNEL CLASSIFICATIONS

To begin, we wish to define the terms *wideband* and *MIMO channel*. This section reviews the theory behind several categories of channels. The section makes the progression from the linear time-invariant SISO channel to the wideband MIMO channel. This leads naturally to the development of a wideband MIMO system model using tensors in Section 2.2.8.

2.2.1 Linear Time-Invariant Channels

We begin by assuming that the wireless channel is linear. Recall that the wireless channel itself is governed by Maxwell's equations, which are inherently linear. The following discussion does not include the effects of nonlinearities in the signal path, such as power amplifiers and mixers. Assuming the channel is linear, we can then treat it as a linear filter. In this way, we can ignore the exact composition of the physical channel and focus on its linear input–output relationship.

Most introductory texts on linear system theory focus on a subset of linear filters: *time-invariant, causal,* and *time-dispersive.* For channels that can be modeled as this category of filtering, the input–output relationship of the channel is determined by the convolution of the input $x(t)$ and the channel's *impulse response* $h(t)$,

$$y(t) = h(t) * x(t), \qquad (2.1)$$

where $*$ denotes convolution. We can also express this relationship in the frequency via the Fourier transform as

$$Y(f) = H(f)X(f), \qquad (2.2)$$

where $y(t) \rightleftharpoons Y(f)$, $x(t) \rightleftharpoons X(f)$, and $h(t) \rightleftharpoons H(f)$. We refer to $H(f)$ as the channel *transfer function.* It is a description of how the channel affects the spectrum of the input signal.

A channel's impulse response completely characterizes its input–output relationship. The implicit assumption is that all energy from all paths in the physical channel combine to form a single response. *Time invariance* implies that $h(t)$ does not change over time. Put differently, a time shift of the input corresponds with the same time shift at the output. A *causal* channel produces an output only after the input has been applied and not before. This is most often the case in physical channels.

A channel is *time-dispersive* when energy at the input at any given time instant is spread out over time at the output. This is illustrated in Fig. 2.2, for $x(t) = \delta(t)$. For an impulsive input, a channel that is not time-dispersive will affect the amplitude and location in time of the impulse, but not its duration. A time-dispersive channel, on the other hand, spreads the energy of the impulse in time.

For time-dispersive channels, $H(f)$ is not constant with respect to f. This in turn means that the channel will attenuate some frequency components of $X(f)$ more than others. Because of this, time-dispersive channels are said to be *frequency-selective.* Communication engineers also refer to the time-dispersive/frequency-selective channel as *wideband.* In the literature, the terms *time-dispersive, frequency-selective,* and

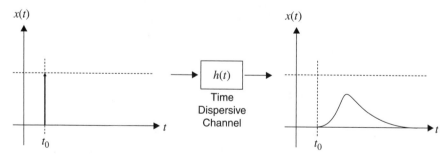

Figure 2.2   The effects of a time-dispersive channel on an impulsive input.

wideband are often used interchangeably. All three terms, however, refer to different facets of the same phenomenon.

In addition to the above effects, real-life channels add noise to the transmitted signal. The noise is usually modeled as a stochastic process. In the following, we consider the case where the noise is *additive*. The resulting model is

$$y(t) = h(t) * x(t) + n(t), \tag{2.3}$$

where $*$ denotes the convolution operator, and $n(t)$ is a stochastic process that embodies the combined effects of all noise sources at the channel output. If $n(t)$ is white and complex Gaussian distributed, then (2.3) is referred to as the *additive white Gaussian noise* (AWGN) channel model, extended to time-dispersive channels. The noise in many physical channels is well approximated by the Gaussian assumption, by virtue of the central limit theorem from probability theory (12).

2.2.2 Time-Invariant Narrowband Channels

We can simplify the input–output relationship (2.3) by assuming that the channel is *not* time-dispersive. In this case, the impulse response does not depend on the delay, and the impulse response is reduced to a single complex constant; that is, $h(t) = h \cdot \delta(t)$. We thus rewrite (2.3) as

$$y(t) = h \cdot x(t) + n(t), \tag{2.4}$$

where we refer to $h \in \mathbb{C}$ as the *complex gain* of the channel. In the case where we have a *narrowband* channel, we do not consider the effects of time dispersion. In the literature, channel models that rely on the narrowband assumption are by far the most common. This is mostly due to their simplicity; narrowband channel models reduce the effects of a physical channel to a single complex number.

2.2.3 Time-Varying Wideband Channels and Bello's Model

If the impulse response is *time-varying* in addition to being time-dispersive, we need a second variable to account for the differences over time. To this end, let the entity that characterizes the input–output relationship of a time-varying, time-dispersive channel be the *kernel function* $k(t, t')$, defined at two times t and t'. Using the kernel function, the input–output relationship can be written as

$$y(t) = \int k(t, t')x(t') \, dt'. \tag{2.5}$$

To bring some physical meaning to $k(t, t')$, we wish to express it in terms of a time t and a *delay* τ. The *input delay-spread* $h_1(t, \tau)$ is defined as the response of the channel at time t to a unit impulse at time $t - \tau$. Substituting $t' = t - \tau$ into (2.5), we get

$$y(t) = \int h_1(t, \tau)x(t - \tau)\,d\tau. \qquad (2.6)$$

Given $h_1(t, \tau)$, it is useful to analyze exactly how rapidly the channel varies in time and quantify how dispersive the channel is.

The Fourier transform is a useful tool to analyze the *frequency content* of a given function. If we take the Fourier transform of $h_1(t, \tau)$ with respect to time t, we can quantify how much the channel varies over time versus delay. If we find the frequency dual with respect to delay τ, we can quantify how dispersive the channel is at a given time. Analytically, the *delay-Doppler spread* $U(v, \tau)$ is defined as

$$U(v, \tau) = \mathop{F}_{t \to v}\{h_1(t, \tau)\} = \int h_1(t, \tau)e^{-j2\pi vt}\,dt, \qquad (2.7)$$

where v is the *Doppler frequency* in hertz. In the literature, $U(v, \tau)$ is often plotted for narrowband channels, or equivalently for a fixed τ. In this case, we speak of the *Doppler spectrum* $U(v)$, which tells us how much the impulse response of a channel changes at a given rate, or the energy at that Doppler frequency.

We can compute the *time-variant transfer function* $T(t, f)$ by taking the Fourier transform of $h_1(t, \tau)$ with respect to τ, viz.,

$$T(t, f) = \mathop{F}_{\tau \to f}\{h_1(t, \tau)\} = \int h_1(t, \tau)e^{-j2\pi f\tau}\,d\tau. \qquad (2.8)$$

The time-variant transfer function characterizes the *frequency selectivity* of the channel at a given instant in time.

The *output Doppler-spread* $G(v, f)$ is found by taking the double Fourier transform of both time and delay simultaneously; that is,

$$G(v, f) = \mathop{F}_{t \to v, \tau \to f}\{h_1(t, \tau)\} = \int\int h_1(t, \tau)e^{-j2\pi vt}e^{-j2\pi f\tau}\,dt\,d\tau. \qquad (2.9)$$

Note that $G(v, f)$ can also be computed from $U(v, \tau)$ and $T(t, f)$; that is,

$$G(v, f) = \mathop{F}_{\tau \to f}\{U(v, \tau)\} \qquad (2.10)$$

and

$$G(v, f) = \mathop{F}_{t \to v}\{T(t, f)\}. \qquad (2.11)$$

Using $G(v, f)$, we can analyze the Doppler spectrum at a fixed frequency, or conversely the channel spectrum at a fixed Doppler frequency.

Channel Spread, Power Delay Profile, and Channel Coherence

So when is a channel considered to be time-variant or wideband? These are simple questions that, unfortunately, do not have simple answers. Whether a channel is time-variant/time-invariant or wideband/narrowband, is a function of the type of signals generated by a system using that channel.

A channel's frequency selectivity is quantified by its span in the delay domain, or its delay spread. As the channel's frequency response becomes more varied, the Fourier relationship defined by Bello's model tells us that an interacting signal will exhibit a broader envelope in the delay domain. In practice, delay spread is the result of received signals from reflectors at large distances versus the line-of-sight (LoS) path between the transmitter and receiver. Increasing the transmit power or receive sensitivity will also increase the apparent delay spread.

We begin by considering the channel for a short-enough period over which it is essentially time-invariant; that is, $h_1(t, \tau)$ does not depend on t and can be expressed as $h_1(\tau)$. Recall that, in general, $h_1(\tau)$ is a complex random function. The *power delay profile* (PDP) is defined as the envelope of $h_1(\tau)$,

$$P(\tau) = |h_1(\tau)|^2. \tag{2.12}$$

The *root mean-square* (RMS) delay spread is related to the statistics of $P(\tau)$. Specifically, the RMS delay spread σ_τ^2 is defined as the second central moment of the PDP (58),

$$\sigma_\tau^2 = \mathrm{E}\{\overline{\tau^2}\} - \mathrm{E}^2\{\overline{\tau}\}, \tag{2.13}$$

where

$$\mathrm{E}\{\overline{\tau^n}\} = \frac{\int \tau^n P(\tau)\, d\tau}{\int P(\tau)\, d\tau}. \tag{2.14}$$

Closely related to the RMS delay spread is the concept of *coherence bandwidth* W_c. Coherence bandwidth is inversely proportional to coherence time. The coherence bandwidth is a statistical measure of the bandwidth over which the channel exhibits approximately equal gain. Conversely, for any two frequency components, f_1 and f_2 are chosen such that $|f_1 - f_2| < W_c$, they will be correlated. The coherence bandwidth is a function of the correlation level and is approximated from the RMS delay spread. For example, if we wish to guarantee that f_1 and f_2 correlate better than 90% of the time, then (59)

$$W_c \approx \frac{1}{50\sigma_\tau}. \tag{2.15}$$

In the literature, the coherence bandwidth is most often defined for the case where we have better than 50% correlation, in which case (60),

$$W_c \approx \frac{1}{5\sigma_\tau}. \tag{2.16}$$

A channel's time variance is quantified by the span of its Doppler spectrum, or its *Doppler spread*; channels that vary faster exhibit a broader frequency range in the Doppler domain. Consider the Doppler spectrum $U(v)$, where v is the Doppler frequency. Channel time-variation can occur due to motion at the transmitter or receiver (or both) or can be caused by scatterer movement in the channel itself. Similar to the RMS delay spread, the RMS Doppler spectrum σ_v^2 is the second central moment of the Doppler spectrum; that is,

$$\sigma_v^2 = \mathrm{E}\{\overline{v^2}\} - \mathrm{E}^2\{\overline{v}\}, \tag{2.17}$$

where

$$\mathrm{E}\{\overline{v^n}\} = \frac{\int v^n P(v)\,dv}{\int P(v)\,dv}. \tag{2.18}$$

In general, a transmitted signal will not be affected appreciably if its baseband-equivalent bandwidth is significantly greater than σ_v^2 (58).

Closely related to the Doppler spread is the *coherence time* T_c. The coherence time is a statistical measure of the time period over which the channel does not change appreciably. Put differently, the channel will affect two signals differently if the period between their arrival times exceeds T_c. The coherence time is inversely proportional to σ_v^2. In the literature, we often see the following rule-of-thumb used to estimate T_c from σ_v^2, (see, for example, Ref. 60):

$$T_c \approx \frac{1}{5\sigma_v}. \tag{2.19}$$

In mobile channels where time-variation can be mainly attributed to movement at the transmitter or receiver, the coherence time is often estimated using the *maximum Doppler shift* f_D, where $f_D = v/\lambda$, v is the velocity of the transmitter or receiver, and λ is the wavelength. In this case, another popular rule-of-thumb for estimating T_c is given as (58)

$$T_c \approx \frac{0.423}{f_D}. \tag{2.20}$$

There are many practical instances when the impulse response does not vary appreciably over a given time. In this case, we say the channel is *static*. There are yet other instances where the impulse response varies slowly over time. In this case, if the observation period T_o is short enough such that the impulse response does not change significantly over T_o, we can also consider it to be constant within the observation period. This is an example of a *slow-fading* channel. In the literature, we often see the terms *static* and *slow-fading* used interchangeably to describe a channel whose impulse response does not vary over time.

In the case of a slow-fading channel, Bello's time-variant model reduces to (2.3). Specifically, if the input delay-spread $h_1(t, \tau)$ does not depend on t, the delay-Doppler spread $U(v, \tau)$ is identically zero, and the time-dependent transfer function $T(t, f)$ depends on f alone. In this case, $T(t, f)$ reduces to $H(f)$, and the input–output relationship is given by (2.3).

2.2.4 The Tapped-Delay Line Model and the Physical Channel

Given the prevalence of digital hardware, it is useful to describe the channel as a function of discrete variables. Consider the case where the wideband channel is time-invariant. We define the discrete function $h[d]$ to be the *complex gain* of the channel at delay d,

$$h[d] = \int h(t)\delta[t - (d - 1)T]\,dt, \tag{2.21}$$

where $\delta(t)$ is the Dirac delta function, $T = 1/2W$, and W is the bandwidth of the channel. In the following, we assume that the channel is causal and of finite length; that is, $d \in \{1, \ldots, D\}$. The *maximum delay* D is determined by the maximum resolvable delay of the system, which, in turn, is determined by the sensitivity of the receiver and is thus a practical consideration.

Rewriting the impulse response as a discrete function greatly simplifies the input–output relationship of the wideband SISO channel. In discrete time, if the input to the channel is $x[i]$ and the output $y[i]$, we can rewrite the input–output relationship for the time-invariant wideband channel as

$$y[i] = \sum_{d=1}^{D} h[d]x[i - (d - 1)] + n[i], \tag{2.22}$$

where $h[d]$ is the set of channel gains for all delays $d = \{1, \ldots, D\}$, and $n[i]$ is a sampled complex-Gaussian noise process. Often, when visualizing $h[d]$, we look at the power profile of $h[d]$ with respect to d and ignore its phase. That is, we often only consider $|h[d]|^2$, which is the discrete *power delay profile* (PDP).

Upon further inspection, (2.22) yields some insight into the composition of physical channels. First, consider the channel structure illustrated in Fig. 2.3. Wideband systems give us the ability to *resolve the channel in time*. This means that, using a wideband system, we can differentiate between signals that arrive at different delays. If the bandwidth of the channel is W, then we can differentiate

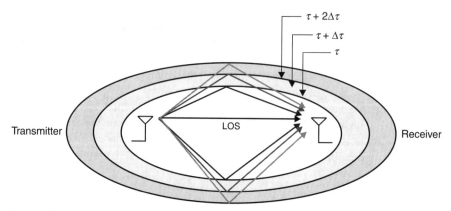

Figure 2.3 Abstraction showing how scatterers in the channel contribute to the overall channel response. Each oval represents a constant-length propagation path from the transmitter to receiver.

between signals arriving at least $\Delta\tau = 1/W$ s apart. The time a signal takes to travel from the transmitter to the receiver depends on its path length. Conceptually, we can divide the channel into a continuum of paths of equal length. These paths necessarily trace out ovals, with the transmitter and receiver at the foci. For a given oval, all paths from the transmitter to any point on the oval, then to the receiver, have equal length, and thus equal propagation delay. The propagation delay to the next resolvable oval is $\Delta\tau$. All energy reflected from scatterers at each oval sums up to form a single response. As we move outward, energy reflected from successive ovals takes longer to get to the receiver. In this way, at the receiver, we observe multiple copies of the transmitted signal, each copy taking progressively longer to reach the receiver. Also, because the distance traveled increases with each oval, the signal reflected off each oval tends to get weaker as we move outward. This model does not consider second- or higher-order reflections, where energy is reflected between scatterers. Indeed, most ray-tracing type models such as Jakes' correlative model (12) are *single-bounce scattering* models in that they assume the majority of the scattered energy arriving at the receiver is due to single-bounce scattering.

The tapped-delay line model, shown in Fig. 2.4, is a diagrammatic representation of (2.22). The delay line produces multiple copies of the channel input at successively longer delays. The time delay between each tap is determined by the bandwidth of the system. Each tap is weighted by a single complex channel gain. In this model, each tap can be viewed as a single SISO channel. The outputs of all taps are summed to produce a single output.

Note that this model is discrete; it assumes that echoes occur at discrete delays, or *delay taps*. This discrete system model is by far the most popular in the literature (see, for example, Ref. 46). This is largely due to the prevalence of digital hardware for measuring and/or computing the channel impulse response. Most channel sounding devices operate in the digital domain and thus sample the PDP at discrete intervals.

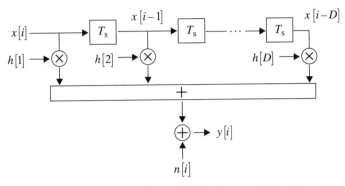

Figure 2.4 A block diagram of the tapped-delay line model for linear time-invariant dispersive channels.

Each delay bin is the sum of all energy received over the duration of that bin. This is analogous to discrete frequency spectra, in which each frequency bin is the sum of all energy over the bandwidth that bin occupies (61).

2.2.5 Narrowband Diversity Channels

Above, we discussed cases where the transmitter and receiver are equipped with a single antenna. In the following, we consider cases where either the transmitter *or* receiver is equipped with multiple antennas. This leads naturally into a discussion of MIMO systems. For simplicity, we begin by considering the narrowband, time-invariant case.

A *multiple-input, single-output* (MISO) system consists of multiple antennas at the transmitter and a single antenna at the receiver. This is also an example of a *transmit-diversity* system. From the above discussion, we know that we require a complex gain to describe the SISO channel from each transmit-antenna to the receive-antenna. The channel response necessarily takes on the form of a vector, with each element in the vector being the complex gain from one transmit-antenna to the receive-antenna. We refer to all possible SISO channels collectively as a *vector channel* or a *path*, where the two terms are used interchangeably in the literature. If we have M_{Tx} transmit-antennas, the input–output relationship for a narrowband time-invariant MISO system becomes

$$y(t) = \mathbf{h}_{\text{MISO}}^T \mathbf{x}(t) + n(t), \tag{2.23}$$

where $\mathbf{h}_{\text{MISO}} = \begin{bmatrix} h_1 & h_2 & \cdots & h_{M_{\text{Tx}}} \end{bmatrix}^T$ is the MISO *channel vector*, or simply *h-vector*. The elements of $\mathbf{h}_{\text{MISO}}$ consist of the complex gains of all possible SISO channels, and the input $\mathbf{x}(t)$ is the *transmit vector*, describing the signal at each transmit-antenna. Note that for M_{Tx} antennas, we have up to M_{Tx} resolvable paths at the transmitter. This is illustrated in Fig. 2.5.

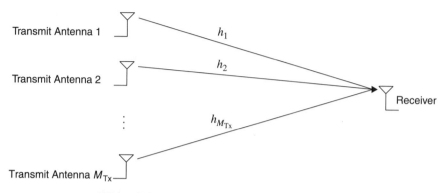

Figure 2.5 Block diagram of the MISO channel.

Conversely, a single-input, multiple-output (SIMO) system comprises a single antenna at the transmit side and M_{Rx} antennas at the receiver. This is also referred to as a *receive-diversity system*. The channel input–output relationship thus takes on the form

$$\mathbf{y}(t) = \mathbf{h}_{SIMO}x(t) + \mathbf{n}(t), \tag{2.24}$$

where $\mathbf{h}_{SIMO} = \begin{bmatrix} h_1 & h_2 & \cdots & h_{M_{Rx}} \end{bmatrix}^T$ is the SIMO h-vector, $\mathbf{n}(t)$ is the receive noise vector, and $\mathbf{y}(t)$ is the *receive vector*. Note that each receive-antenna is associated with an additive complex-Gaussian distributed noise process that is independent of any other receive-antenna. The SIMO channel is illustrated in Fig. 2.6.

Depending on the composition of the physical channel, the signal at a given receive-antenna can be independent of all others, or it can correlate in some way. As we will discuss in Section 2.6.2, correlation between signals at different antennas allows us to deduce the location of scatterers in the physical channel.

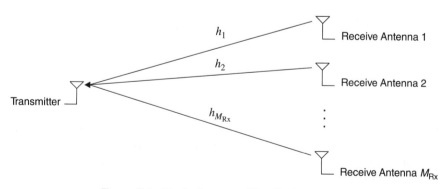

Figure 2.6 Block diagram of the SIMO channel.

2.2.6 The Narrowband MIMO Channel

Consider the case where both the transmitter *and* receiver are equipped with multiple antennas. This introduces the ability to resolve the spatial structure at both link-ends, and thus gives us an added degree of freedom. When a system is equipped with multiple antennas at the transmitter and receiver, a MIMO channel separates them. To simplify the discussion, the MIMO channel can be viewed as a collection of many SISO channels. One SISO channel exists between each transmit-receive pair. As with the diversity case above, each SISO channel is often referred to as a *path* or *parallel channel*. To be clear, we refer to a *single* MIMO channel as being composed of many *paths*. This is illustrated in Fig. 2.7 for the 2×2 case.

For the narrowband MIMO channel, we use a complex gain to characterize each path. Thus, we express all complex gains as a matrix **H**, where

$$\mathbf{H} = \begin{pmatrix} h_{11} & \cdots & h_{1M_{\mathrm{Tx}}} \\ \vdots & \ddots & \vdots \\ h_{M_{\mathrm{Rx}}1} & \cdots & h_{M_{\mathrm{Rx}}M_{\mathrm{Tx}}} \end{pmatrix}. \tag{2.25}$$

We refer to **H** as the *H-matrix*, and thus to the MIMO channel as the *matrix channel*. Each element of **H** is the complex gain h_{mn} between receive-antenna m, and transmit-antenna n, for $m \in \{1, \ldots, M_{\mathrm{Rx}}\}$ and $n \in \{1, \ldots, M_{\mathrm{Tx}}\}$.

Spatial Multiplexing versus Spatial Diversity   Here, we briefly comment on the importance of MIMO channels. One of the major advantages of MIMO systems is their ability to transmit more than one stream of information at the same time and within the same bandwidth, thus making a MIMO channel more bandwidth-efficient than a SISO channel. This is best illustrated with an example. Consider the 2×2 MIMO channel shown in Fig. 2.8.

At the transmitter, antenna $n = \{1, 2\}$ transmits signal $s_n(t)$. The transmit antennas are separated far enough in space such that they both "see" different paths to the receive-antennas. Both signals are allowed to co-mingle in the channel. At the receiver, each receive-antenna captures a different signal. Each signal is an independently faded version of $s_1(t) + s_2(t)$. Through the use of *space–time codes*, if the paths characterized by h_{mn} are statistically distinct (i.e., they *fade independently*), it is possible to separate $s_1(t)$ and $s_2(t)$ using the signals from both receive-antennas.

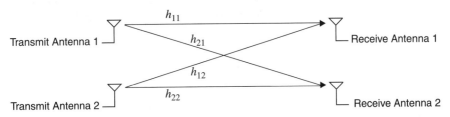

Figure 2.7   A 2×2 MIMO channel.

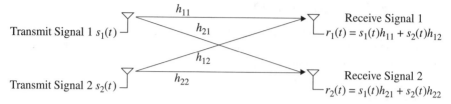

Figure 2.8 A 2×2 MIMO channel, highlighting how transmitted signals are affected in the channel.

This implies that we can increase the overall capacity by using multiple antennas (see, for example, Refs. 17, 22, 38). This is an example of *spatial multiplexing*. The increased capacity due to spatial multiplexing is referred to as the *multiplexing gain*. We measure the distinctness of a path by measuring its *statistical independence* from other paths. In reality, the capacity of a MIMO channel is limited by the number and strength of independent paths. The number of independent paths, in turn, is largely determined by the spatial structure of the physical channel.

MIMO systems have the added benefit that they can strike a trade-off between multiplexing and diversity gain. By changing the way in which information is coded at the transmitter, we can either send multiple copies of the same message on different antennas (diversity gain), send different messages on different antennas (multiplexing gain), or strike a balance somewhere in between. If all paths from the transmitter to receiver are faded, or if they exhibit a large variance (and are thus unreliable), we should use diversity to maximize the average capacity. If most paths are good, that is, they exhibit a high signal-to-noise ratio (SNR), and are relatively static, then we should use spatial multiplexing to maximize the average capacity. This is referred to as the *diversity-multiplexing trade-off* and is currently the subject of intense study (see, for example, Refs. 24, 62). The diversity-multiplexing trade-off is an information-theoretic topic and lies outside the scope of this text.

2.2.7 The Wideband MIMO Channel

The above definition of the H-matrix has been extended to include wideband MIMO channels. In the literature, the wideband MIMO channel is most often assumed to be time-invariant and is expressed using discrete rather than continuous variables (see, for example, Ref. 63). Thus, the wideband SISO channel between each transmit-receive pair can be characterized using the complex channel gain $h_{mn}[d]$. For all possible transmit-receive pairs, we may extend (2.25) to the wideband case,

$$\mathbf{H}[d] = \begin{pmatrix} h_{11}[d] & \cdots & h_{1M_{Tx}}[d] \\ \vdots & \ddots & \vdots \\ h_{M_{Rx}1}[d] & \cdots & h_{M_{Rx}M_{Tx}}[d] \end{pmatrix}. \tag{2.26}$$

The elements of the *wideband H-matrix* $\mathbf{H}[d]$ are the complex channel gains $h_{mn}[d]$. Each $h_{mn}[d]$ describes the complex gain between receive-antenna m, transmit-antenna n, at delay d.

For the wideband MIMO case, the input–output relationship can be written as

$$\mathbf{y}[i] = \sum_{d=1}^{D} \mathbf{H}[d]\mathbf{x}[i - (d-1)] + \mathbf{n}[i], \tag{2.27}$$

where $\mathbf{x}[i]$ is the $M_{\mathrm{Tx}} \times 1$ channel input for all transmitters at index i, $\mathbf{n}[i]$ is the $M_{\mathrm{Rx}} \times 1$ noise vector, and $\mathbf{y}[i]$ is the $M_{\mathrm{Rx}} \times 1$ receive vector.

2.2.8 The Wideband MIMO Channel Recast Using Tensors

In the following section, we recast the complex channel gains in a wideband MIMO channel as a third-order tensor, which we call the *H-tensor*. This represents a small but significant step; recasting the complex channel gains as a tensor allows the use of tensor decomposition in modeling the wideband MIMO channel. Using the H-tensor, we recast the wideband MIMO system model given in (2.27). Appendix A presents a brief introduction to tensor calculus and introduces some notation and operations used here.

Consider the wideband H-matrix $\mathbf{H}[d]$, which was first defined in Section 2.2.7. To characterize the wideband MIMO channel, we need all the elements $\mathbf{H}[d]$ for all $m \in \{1, \ldots, M_{\mathrm{Rx}}\}$, $n \in \{1, \ldots, M_{\mathrm{Tx}}\}$, and $d \in \{1, \ldots, D\}$. Essentially, this means that $\mathbf{H}[d]$ is a three-dimensional entity, with each element necessarily addressed by the triplet (m, n, d). As such, it becomes awkward to define channel operations using $\mathbf{H}[d]$ and matrix algebra. If we recast $\mathbf{H}[d]$ as a third-order tensor, we can use mul-tilinear algebra to analyze the wideband MIMO channel in a rather compact manner.

We define the *wideband H-tensor* $\mathcal{H} \in \mathbb{C}^{M_{\mathrm{Rx}} \times M_{\mathrm{Tx}} \times D}$ to be a tensor whose element h_{mnd} is the complex gain of the channel between receiver m, transmitter n at delay d, for $m \in \{1, \ldots, M_{\mathrm{Rx}}\}$, $n \in \{1, \ldots, M_{\mathrm{Tx}}\}$, and $d \in \{1, \ldots, D\}$. The structure of the H-tensor is illustrated in Fig. 2.9.

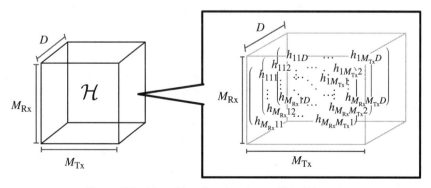

Figure 2.9 Visualizing the structure of the H-tensor.

In this way, the wideband H-matrix $\mathbf{H}[d]$ consists of the subset of elements of $\mathcal{H}$ corresponding with delay tap d. We can rewrite (2.26) in terms of the H-tensor elements h_{mnd} as

$$\mathbf{H}[d] = \begin{pmatrix} h_{11d} & \cdots & h_{1M_{\text{Tx}}d} \\ \vdots & \ddots & \vdots \\ h_{M_{\text{Rx}}1d} & \cdots & h_{M_{\text{Rx}}M_{\text{Tx}}d} \end{pmatrix}. \tag{2.28}$$

Note that when $D = 1$, the H-tensor reduces to the familiar H-matrix $\mathbf{H} \in \mathbb{C}^{M_{\text{Rx}} \times M_{\text{Tx}}}$.

Similarly, from (2.27), the channel input vector $\mathbf{x}[d]$ is a two-dimensional quantity. We can map the elements of the channel input vector $\mathbf{x}[d]$ for $d = \{0, -1, \ldots, -(D-1)\}$ to the matrix element x_{nd}; for example,

$$\mathbf{x}[d] = [x_{1d} \quad x_{2d} \quad \cdots \quad x_{M_{\text{Tx}}d}]^T. \tag{2.29}$$

Using the tensor summation convention (see Appendix A), we can recast (2.27) as

$$y_m = h_{mnd}x_{nd} + n_m, \tag{2.30}$$

where $\mathbf{n} \in \mathbb{C}^{M_{\text{Rx}} \times 1}$ is the noise vector, and $\mathbf{y} \in \mathbb{C}^{M_{\text{Rx}} \times 1}$ is the channel input vector. In the above model, we only consider the output for a single time instant, and thus drop the index i.

2.3 SUMMARY OF CHANNEL CLASSIFICATIONS

In the previous sections, we have classified the set of all possible wireless channels according to time variation, time dispersion, and their topology; for example, if the channel is SISO, MIMO, and so forth. Table 2.1 presents a summary of these classes. Although this is in no way a complete characterization of all possible wireless channels, it covers those most often encountered in the literature.

2.4 SECOND-ORDER STATISTICS OF MULTIPLE ANTENNA CHANNELS

The following section defines correlation in multiantenna channels. We discuss how the correlation between paths in a multiantenna channel characterizes its spatial structure. The section begins by discussing correlation in vector channels. This leads to a discussion of correlation in narrowband MIMO channels. We then discuss how the eigenvalue decomposition (EVD) of the correlation matrix reveals more about the statistical behavior of scatterers in the channel. Section 2.5 presents a formulation for the correlation in wideband MIMO channels using tensor calculus. Section 2.6 discusses how channel correlation can be used to characterize the spatial structure of a given channel. In addition, channel correlation is the basis for correlative channel models, as discussed in Chapter 3.

Table 2.1 Channel classification according to time variation, time dispersion, and channel topology

Time Variation	Time Dispersion	Channel Topology	Characterization	Comments
Time invariant	Time dispersive (a.k.a. wideband, frequency selective)	SISO	$h(t)$	Impulse response depends on t, but $h(t)$ does not change over time. Continuous variable.
Time invariant	Not time dispersive (a.k.a. narrowband, frequency flat fading)	SISO	h	Constant, does not depend on t. Does not change over time.
Time variant	Time dispersive	SISO	$h(t,\tau)$	Impulse response depends on t, and $h(t)$ changes over time; the delay τ is needed to describe this change. Continuous variable.
Time invariant	Not time dispersive	SIMO, MISO	$\mathbf{h}_{\mathrm{MISO}} = [h_1 \quad h_2 \quad \cdots \quad h_{M_{\mathrm{Tx}}}]^T$ $\mathbf{h}_{\mathrm{SIMO}} = [h_1 \quad h_2 \quad \cdots \quad h_{M_{\mathrm{Rx}}}]^T$	Vector channel with M_{Tx}, M_{Rx} constant complex coefficients, respectively.
Time invariant	Not time dispersive	MIMO	$\mathbf{H} = \begin{pmatrix} h_{11} & \cdots & h_{1M_{\mathrm{Tx}}} \\ \vdots & \ddots & \vdots \\ h_{M_{\mathrm{Rx}}1} & \cdots & h_{M_{\mathrm{Rx}}M_{\mathrm{Tx}}} \end{pmatrix}$	H-matrix, consisting of $M_{\mathrm{Rx}} \times M_{\mathrm{Tx}}$ constant complex coefficients.
Time invariant	Time dispersive	MIMO	$\mathbf{H}[d] = \begin{pmatrix} h_{11}[d] & \cdots & h_{1M_{\mathrm{Tx}}}[d] \\ \vdots & \ddots & \vdots \\ h_{M_{\mathrm{Rx}}1}[d] & \cdots & h_{M_{\mathrm{Rx}}M_{\mathrm{Tx}}}[d] \end{pmatrix}$ $\mathcal{H} \in \mathbb{C}^{M_{\mathrm{Rx}} \times M_{\mathrm{Tx}} \times D}$	"Indexed matrix," a discrete sampling of each of $M_{\mathrm{Rx}} \times M_{\mathrm{Tx}} \times D$ time-dispersive channels represented in matrix form. Unweildy. H-tensor, discrete represen-ation of the time-dispersive MIMO channel. Makes it easier to do linear operations.

2.4.1 Second-Order Statistics of the Vector Channel

For many practical cases, the complex gains in a vector channel can be modeled as complex-Gaussian distributed random variables. This means that the distribution of each complex gain can be fully characterized by the first two moments, namely its *mean* and *variance*. If the mean and variance of the channel do not change over time, then the channel is said to be *wide-sense stationary* (WSS). For a WSS channel, we can compute the correlation between paths in the vector channel. Consider the vector channel characterized by $\mathbf{h} = [h_1 \ h_2 \ \cdots \ h_M]^T$. Let r_{mn} be the complex correlation between complex gains h_m and h_n for $m, n \in \{1, \ldots, M\}$; that is,

$$r_{mn} = \mathrm{E}\{h_m h_n^*\} \tag{2.31}$$

where $\mathrm{E}\{\cdot\}$ is the expectation operator, and $(\cdot)^*$ is the complex conjugate. The correlation between two channel coefficients represents the average energy coupled between them.

We can compute the correlation between all paths by taking the statistical expectation of the complex outer product of $\mathbf{h}$ with itself; that is,

$$\mathbf{R}_h = \mathrm{E}\{\mathbf{h}\mathbf{h}^H\}$$

$$= \mathrm{E}\left\{ \begin{pmatrix} h_1 h_1^* & \cdots & h_1 h_M^* \\ \vdots & \ddots & \vdots \\ h_M h_1^* & \cdots & h_M h_M^* \end{pmatrix} \right\}. \tag{2.32}$$

We refer to $\mathbf{R}_h$ as the *vector correlation matrix*. Because it is computed as the complex outer product of the same vector, the vector correlation matrix is *Hermitian*, that is $r_{mn} = r_{nm}^*$, or equivalently $\mathbf{R}_h = \mathbf{R}_h^H$, where $(\cdot)^H$ is the Hermitian transpose. The diagonal elements r_{mm} are real-valued and represent the average power of the mth channel.

2.4.2 Second-Order Statistics of the Narrowband MIMO Channel

The correlation between all paths in a matrix channel can be defined in much the same way as for a vector channel. Given an H-matrix, the *narrowband correlation matrix* $\mathbf{R}_H$ is defined as

$$\mathbf{R}_H = \mathrm{E}\{\mathrm{vec}(\mathbf{H})\mathrm{vec}^H(\mathbf{H})\}, \tag{2.33}$$

where $\mathrm{vec}(\mathbf{H})$ is the mapping of all elements in $\mathbf{H}$ such that all columns in $\mathbf{H}$ are stacked on top of each other to form a vector. Written explicitly,

$$\mathrm{vec}(\mathbf{H}) = [h_{11} \ \cdots \ h_{M_{\mathrm{Rx}}1} \ h_{12} \ \cdots \ h_{M_{\mathrm{Rx}}2} \ \cdots \ h_{M_{\mathrm{Rx}}M_{\mathrm{Tx}}}]^T. \tag{2.34}$$

The elements of $\mathbf{R}_H$ can be viewed as the correlation between all possible paths; that is,

$$\mathbf{R}_H =$$

$$\mathrm{E}\left\{\begin{pmatrix} h_{11}h_{11}^* & \cdots & h_{11}h_{M_{\mathrm{Rx}}1}^* & h_{11}h_{12}^* & \cdots & h_{11}h_{M_{\mathrm{Rx}}2}^* & \cdots & h_{11}h_{M_{\mathrm{Rx}}M_{\mathrm{Tx}}}^* \\ \vdots & \ddots & \vdots & \vdots & & \vdots & & \vdots \\ h_{M_{\mathrm{Rx}}1}h_{11}^* & \cdots & h_{M_{\mathrm{Rx}}1}h_{M_{\mathrm{Rx}}1}^* & h_{M_{\mathrm{Rx}}1}h_{12}^* & \cdots & h_{M_{\mathrm{Rx}}1}h_{M_{\mathrm{Rx}}2}^* & \cdots & h_{M_{\mathrm{Rx}}1}h_{M_{\mathrm{Rx}}M_{\mathrm{Tx}}}^* \\ h_{12}h_{11}^* & \cdots & h_{12}h_{M_{\mathrm{Rx}}1}^* & h_{12}h_{12}^* & \cdots & h_{12}h_{M_{\mathrm{Rx}}2}^* & \cdots & h_{12}h_{M_{\mathrm{Rx}}M_{\mathrm{Tx}}}^* \\ \vdots & & \vdots & \vdots & \ddots & \vdots & & \vdots \\ h_{M_{\mathrm{Rx}}2}h_{11}^* & \cdots & h_{M_{\mathrm{Rx}}2}h_{M_{\mathrm{Rx}}1}^* & h_{M_{\mathrm{Rx}}2}h_{12}^* & \cdots & h_{M_{\mathrm{Rx}}2}h_{M_{\mathrm{Rx}}2}^* & \cdots & h_{M_{\mathrm{Rx}}2}h_{M_{\mathrm{Rx}}M_{\mathrm{Tx}}}^* \\ \vdots & & \vdots & \vdots & & \vdots & \ddots & \vdots \\ h_{M_{\mathrm{Rx}}M_{\mathrm{Tx}}}h_{11}^* & \cdots & h_{M_{\mathrm{Rx}}M_{\mathrm{Tx}}}h_{M_{\mathrm{Rx}}1}^* & h_{M_{\mathrm{Rx}}M_{\mathrm{Tx}}}h_{12}^* & \cdots & h_{M_{\mathrm{Rx}}M_{\mathrm{Tx}}}h_{M_{\mathrm{Rx}}2}^* & \cdots & h_{M_{\mathrm{Rx}}M_{\mathrm{Tx}}}h_{M_{\mathrm{Rx}}M_{\mathrm{Tx}}}^* \end{pmatrix}\right\}.$$

$$(2.35)$$

The narrowband correlation matrix shares many of the same properties as the vector correlation matrix $\mathbf{R}_h$, with one important difference; the elements of $\mathbf{R}_H$ are mapped from a fourth-order tensor. Let the elements of $\mathbf{R}_H$ be r_{mnpq}, where

$$r_{mnpq} = \mathrm{E}\{h_{mn}h_{pq}\}, \qquad (2.36)$$

for m, $p \in \{1, \ldots, M_{\mathrm{Rx}}\}$ and n, $q \in \{1, \ldots, M_{\mathrm{Tx}}\}$. As for the vector case, $\mathbf{R}_H$ is Hermitian symmetric. The diagonal elements of $\mathbf{R}_H$, r_{mnmn} are real-valued and represent the average power in h_{mn}. Note that unlike the vector case, we need a quadruplet to address a single element in $\mathbf{R}_H$ properly. Put differently, given h_{mn} and h_{pq}, we need four indices, namely (m, n, p, q), to address a single r_{mnpq}. By vectorizing $\mathbf{H}$ and forcing the MIMO channel correlation to be a matrix, we lose some of the structure.

Eigenvalue Decomposition of the Narrowband Correlation Matrix

Consider the hypothetical channel depicted in Fig. 2.10. Here, there are four resolvable scatterers $s_1 - s_4$, each one associated with a distinct single-bounce path from the transmitter to receiver.

Each path is associated with its own complex gain, labeled h_1 to h_4. Each channel coefficient is a complex-Gaussian random variable. If the correlation between channels $\mathrm{E}\{h_1h_2^*\}$ is close to 1, then we say that the *fading* between paths 1 and 2 is strongly correlated. Equivalently, we say that *the two scatterers s_1 and s_2 are strongly correlated*. In a similar fashion, let $\mathrm{E}\{h_3h_4^*\} = 1$. Let there be no other significant correlation. In this example, we consider $\{s_1, s_2\}$ and $\{s_3, s_4\}$ to be *correlated scatterers*.

In Ref. 51, Weichselberger provides a detailed discussion of the EVD of $\mathbf{R}_H$ and how it can be used to distinguish between correlated and uncorrelated scatterers. The following summarizes some of his findings.

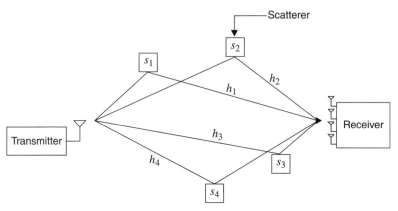

Figure 2.10 Example channel in which four resolvable scatterers reflect transmitted energy toward the receiver. Each discrete path is associated with a single complex gain h_i for $i = \{1, \ldots, 4\}$.

Expressing $\mathbf{R}_H$ via its EVD, we have

$$\mathbf{R}_H = \sum_{k=1}^{M_{\mathrm{Rx}} M_{\mathrm{Tx}}} \lambda_k \mathbf{u}_k \mathbf{u}_k^H, \tag{2.37}$$

where there are up to $M_{\mathrm{Rx}} \times M_{\mathrm{Tx}}$ significant eigenvalues λ_k, and $\mathbf{u}_k$ is the kth eigenvector, associated with the kth eigenvalues λ_k. In a theoretical context, the term *significant* denotes any nonzero eigenvalue. In a practical context, the term *significant* is often determined by the minimum sensitivity of the receiver; all eigenvalues whose power falls below a threshold are most often considered to be zero. The threshold in turn is determined by the minimum sensitivity of the receiver.

From Ref. 64, any given λ_k is associated with the average power of the MIMO channel defined by a group of *correlated* scatterers. Any two eigenvalues λ_k, λ_ℓ for $k \neq \ell$ are associated with groups of scatterers that *are not* correlated. Thus, by taking the EVD of $\mathbf{R}_H$, we can separate the channel into groups of correlated scatterers. Each group of scatterers does not correlate with any other group.

The number of scatterers in each group can be found by recasting the eigenvectors $\mathbf{u}_k$ as matrices. Specifically, consider the mapping of the elements of $\mathbf{u}_k$ to an $M_{\mathrm{Rx}} \times M_{\mathrm{Tx}}$ matrix $\mathbf{U}_k$ such that

$$\mathbf{U}_k = \mathrm{unvec}(\mathbf{u}_k). \tag{2.38}$$

$\mathbf{U}_k$ is referred to as the kth *eigenbasis* of the channel. The *singular value decomposition* (SVD) can be used to further decompose each eigenbasis as

$$\mathbf{U}_k = \sum_{\ell=1}^{M_{\min}} \lambda'[k, \ell] \mathbf{u}_{\mathrm{Rx}, k, \ell} \mathbf{u}_{\mathrm{Tx}, k, \ell}^T, \tag{2.39}$$

where $\lambda'[k, \ell]$ is the ℓth singular value of the kth eigenbasis, $\mathbf{u}_{\mathrm{Rx},k,\ell}$ is the ℓth left singular vector of the kth eigenbasis, and $\mathbf{u}_{\mathrm{Tx},k,\ell}$ is the ℓth right singular vector of the kth eigenbasis. We refer to $\mathbf{u}_{\mathrm{Rx},k,\ell}$ and $\mathbf{u}_{\mathrm{Tx},k,\ell}$ collectively as the *submodes* of the channel. The number of nonzero singular values $\lambda'[k, \ell]$ determines the number of correlated scatterers represented by $\mathbf{U}_k$. In this way, we can use the EVD of the channel correlation to distinguish between correlated and uncorrelated scatterers in a given narrowband MIMO channel.

2.5 SECOND-ORDER STATISTICS OF THE WIDEBAND MIMO CHANNEL

In this section, tensor calculus is used to present a formulation for correlation in a wideband MIMO channel. Specifically, we define two quantities, the *wideband correlation tensor* and the *wideband correlation matrix*. Defining correlation in this way reveals more about the structure of the wideband MIMO channel and leads to the structured model, presented in Section 3.6.

We define the wideband correlation tensor $\mathcal{R} \in \mathbb{C}^{M_{\mathrm{Rx}} \times M_{\mathrm{Tx}} \times D \times M_{\mathrm{Rx}} \times M_{\mathrm{Tx}} \times D}$ to be the expected value of the outer product of the H-tensor with its complex conjugate; that is,

$$\mathcal{R} - \mathrm{E}\{\mathcal{H} \circ \mathcal{H}^*\}, \tag{2.40}$$

where $\circ$ denotes the tensor outer product, defined in Appendix A. The elements of the wideband correlation tensor are computed as

$$r_{mndpqr} \triangleq \mathrm{E}\{h_{mnd}h_{pqr}^*\} \tag{2.41}$$

for $p = \{1, \ldots, M_{\mathrm{Rx}}\}$, $q = \{1, \ldots, M_{\mathrm{Tx}}\}$, and $r = \{1, \ldots, D\}$. Here, we define the conjugate of all elements in $\mathcal{H}$ to be $(\mathcal{H}^*)_{mnd} = h_{mnd}^*$.

To make this definition more clear and to relate the result to established engineering convention, we define wideband correlation as a matrix by first mapping the elements h_{mnd} to a vector and then computing the expectation of the outer product of that vector. The resulting equation will resemble those for $\mathbf{R}_h$ (2.32) and $\mathbf{R}_H$ (2.33). First, we define $\mathbf{h}_{\mathrm{WB}} = \mathrm{vec}(\mathcal{H})$ to be the mapping of all h_{mnd} to a column vector; that is,

$$\mathbf{h}_{\mathrm{WB}} \triangleq \mathrm{vec}(\mathcal{H})$$
$$\triangleq [h_{111} \ h_{211} \ \cdots \ h_{M_{\mathrm{Rx}}11} \ h_{121} \ \cdots \ h_{M_{\mathrm{Rx}}21} \ \cdots \ h_{M_{\mathrm{Rx}}M_{\mathrm{Tx}}1} \ h_{112} \ \cdots \ h_{M_{\mathrm{Rx}}M_{\mathrm{Tx}}D}]^T, \tag{2.42}$$

where the superscript T denotes the transpose operator. In this way, the $\mathrm{vec}(\cdot)$ [and the reverse $\mathrm{unvec}(\cdot)$] operators can be seen as a mapping of a third-order tensor to a vector and vice versa.

Next, we define the $M_{Rx}M_{Tx}D \times M_{Rx}M_{Tx}D$ wideband correlation matrix $\mathbf{R}_{WB,H}$ as the expected value of the outer product of $\mathbf{h}_{WB}$:

$$\mathbf{R}_{WB,H} = E\{\text{vec}(\mathcal{H})\text{vec}^H(\mathcal{H})\}$$

$$= E\left\{\begin{pmatrix} h_{111}h_{111}^* & \cdots & h_{111}h_{M_{Rx}11}^* & h_{111}h_{121}^* & \cdots & h_{111}h_{M_{Rx}21}^* & \cdots & h_{111}h_{M_{Rx}M_{Tx}D}^* \\ \vdots & \ddots & \vdots & \vdots & & \vdots & & \vdots \\ h_{M_{Rx}11}h_{111}^* & \cdots & h_{M_{Rx}11}h_{M_{Rx}11}^* & h_{M_{Rx}11}h_{121}^* & \cdots & h_{M_{Rx}11}h_{M_{Rx}21}^* & \cdots & h_{M_{Rx}11}h_{M_{Rx}M_{Tx}D}^* \\ h_{121}h_{111}^* & \cdots & h_{121}^*h_{M_{Rx}11}^* & h_{121}^*h_{121}^* & \cdots & h_{121}^*h_{M_{Rx}21}^* & \cdots & h_{121}^*h_{M_{Rx}M_{Tx}D}^* \\ \vdots & & \vdots & \vdots & \ddots & \vdots & & \vdots \\ h_{M_{Rx}21}h_{111}^* & \cdots & h_{M_{Rx}21}h_{M_{Rx}11}^* & h_{M_{Rx}21}h_{121}^* & \cdots & h_{M_{Rx}21}h_{M_{Rx}21}^* & \cdots & h_{M_{Rx}21}h_{M_{Rx}M_{Tx}D}^* \\ \vdots & & \vdots & \vdots & & \vdots & \ddots & \vdots \\ h_{M_{Rx}M_{Tx}D}h_{111}^* & \cdots & h_{M_{Rx}M_{Tx}D}h_{M_{Rx}11}^* & h_{M_{Rx}M_{Tx}D}h_{121}^* & \cdots & h_{M_{Rx}M_{Tx}D}h_{M_{Rx}21}^* & \cdots & h_{M_{Rx}M_{Tx}D}h_{M_{Rx}M_{Tx}D}^* \end{pmatrix}\right\}.$$

$$(2.43)$$

Note that $\mathbf{R}_{WB,H}$ is only one of the many possible mappings of the wideband correlation tensor $\mathcal{R}$ to a matrix. This particular mapping has some nice properties. By way of explanation, we can recast $\mathbf{R}_{WB,H}$ using the wideband H-matrix $\mathbf{H}[d]$ as

$$\mathbf{R}_{WB,H} = E\{\text{vec}(\mathcal{H})\text{vec}^H(\mathcal{H})\}$$

$$= \begin{pmatrix} \overbrace{E\{\text{vec}(\mathbf{H}[1])\text{vec}^H(\mathbf{H}[1])\}}^{M_{Rx}M_{Tx} \times M_{Rx}M_{Tx}} & \cdots & E\{\text{vec}(\mathbf{H}[1])\text{vec}^H(\mathbf{H}[D])\} \\ \vdots & \ddots & \vdots \\ E\{\text{vec}(\mathbf{H}[D])\text{vec}^H(\mathbf{H}[1])\} & \cdots & E\{\text{vec}(\mathbf{H}[D])\text{vec}^H(\mathbf{H}[D])\} \end{pmatrix}$$

$$\triangleq \begin{pmatrix} \mathbf{R}_{WB,H[1]H[1]} & \cdots & \mathbf{R}_{WB,H[1]H[D]} \\ \vdots & \ddots & \vdots \\ \mathbf{R}_{WB,H[D]H[1]} & \cdots & \mathbf{R}_{WB,H[D]H[D]} \end{pmatrix}. \qquad (2.44)$$

Because $\mathbf{R}_{WB,H}$ is related to the outer product of the same vector, it is Hermitian symmetric and thus lends itself to EVD. The diagonal blocks of $\mathbf{R}_{WB,H}$ are also Hermitian symmetric, but the off-diagonal blocks are only symmetric to their opposites across the diagonal. That is to say, $\mathbf{R}_{WB,H[d]H[d]} = \mathbf{R}_{WB,H[d]H[d]}^H$ and $\mathbf{R}_{WB,H[c]H[d]} = \mathbf{R}_{WB,H[d]H[c]}^H$, where the superscript H denotes the Hermitian transpose.

For real-life channels, $\mathbf{R}_{WB,H}$ tends to be block diagonal. Qualitatively, consider the oval of scatterers model first shown in Fig. 2.3. Each discrete tap in the PDP represents the sum of energy from all scatterers in successively farther ovals. For many practical channels, the scatterers in each oval are far enough away that they can be considered uncorrelated. In this case, the cross-correlation blocks $\mathbf{R}_{WB,H[c]H[d]}$ for $c \neq d$ tend to zero. The diagonal blocks $\mathbf{R}_{WB,H[d]H[d]}$ are similar to the original narrowband correlation matrix $\mathbf{R}_{WB,H}$ for a given d. As the bandwidth of a system goes up, however, the distance between ovals gets proportionally smaller, and we gain spatial resolution. At some point, it can be expected that the cross-correlation blocks will be significant contributors to the wideband correlation matrix. The above representation covers both extremes.

2.5.1 Eigenvalue Decomposition of the Wideband Correlation Matrix

This section extends the ideas given by Weichselberger in Ref. 51 to the EVD of $\mathbf{R}_{WB,H}$. Because the wideband correlation matrix is Hermitian symmetric, we can apply the EVD directly and thus write

$$\mathbf{R}_{WB,H} = \sum_{k=1}^{M_{Rx}M_{Tx}D} \lambda_{WB,k} \mathbf{u}_{WB,k} \mathbf{u}_{WB,k}^H, \tag{2.45}$$

where $\lambda_{WB,k}$ and $\mathbf{u}_{WB,k}$ are the kth eigenvalues and eigenvectors of the wideband correlation matrix, respectively. Note that the structure of the kth eigenvector follows from the structure of $\mathbf{h}_{WB}$ in (2.42). That is, the structure of the kth eigenbasis follows

$$u_{WB,k} = \begin{bmatrix} u_{111k} & u_{211k} & \cdots & u_{M_{Rx}11k} & u_{121k} & \cdots & u_{M_{Rx}21k} & \cdots & u_{M_{Rx}M_{Tx}1k} & \cdots & u_{M_{Rx}M_{Tx}Dk} \end{bmatrix}^T, \tag{2.46}$$

where u_{mndk} are the (fourth-order) tensor elements of $\mathbf{u}_{WB,k}$.

We can reveal more about the structure of the channel by recasting $\mathbf{u}_{WB,k}$ as a matrix and looking at its SVD. To this end, recasting each eigenvector as a $M_{Rx}M_{Tx} \times D$ matrix leads to

$$\begin{aligned}
\mathbf{U}_{WB,k} &= \mathrm{unvec}(\mathbf{u}_{WB,k}) \\
&= \begin{pmatrix}
u_{111k} & u_{112k} & \cdots & u_{11Dk} \\
\vdots & \vdots & & \vdots \\
u_{M_{Rx}11k} & u_{M_{Rx}12k} & \cdots & u_{M_{Rx}1Dk} \\
u_{121k} & u_{122k} & \cdots & u_{12Dk} \\
\vdots & \vdots & & \vdots \\
u_{M_{Rx}21k} & u_{M_{Rx}22k} & \cdots & u_{M_{Rx}2Dk} \\
\vdots & \vdots & & \vdots \\
u_{M_{Rx}M_{Tx}1k} & u_{M_{Rx}M_{Tx}2k} & \cdots & u_{M_{Rx}M_{Tx}Dk}
\end{pmatrix}.
\end{aligned} \tag{2.47}$$

where $\mathbf{U}_{WB,k}$ is the kth wideband eigenbasis. As with the narrowband case, $\mathbf{U}_{WB,k}$ is the basis for a set of correlated scatterers, with fading coefficient $\lambda_{WB,k}$. Each eigenbasis can be further reduced using the SVD, as shown by

$$\mathbf{U}_{\mathrm{WB},k} = \sum_{l=1}^{M_{\mathrm{min}}} \lambda'[k, \ell] \mathbf{u}_{\mathrm{RxTx},k,\ell} \mathbf{u}_{D,k,\ell}^T, \tag{2.48}$$

where $M_{\mathrm{min}} = \min(M_{\mathrm{Rx}} M_{\mathrm{Tx}}, D)$, $\mathbf{u}_{\mathrm{RxTx},k,\ell}$, and $\mathbf{u}_{D,k,\ell}^T$, are the lth singular value, left singular vector, and right singular vector of the kth wideband eigenbasis, respectively. Note that the above formulation for $\mathbf{U}_{\mathrm{WB},k}$ is not the only way to map the elements of $\mathbf{U}_{\mathrm{WB},k}$ to a matrix. However, this formulation results in a more interesting SVD. The left singular vectors $\mathbf{u}_{\mathrm{RxTx},k,\ell}$ describe the transmit-receive pair space, and the right singular vectors $\mathbf{u}_{D,k,\ell}^T$ describe the delay domain space. In general, the rank of each eigenbasis is the number of correlated (resolvable) scatterers for that basis. The difference here is that the scatterers can be from any oval in Fig. 2.3. If the ovals are well spaced (low system bandwidth), then it is less likely that scatterers from different ovals will show up in the same eigenbasis. As the system bandwidth increases, so should the likelihood of correlated scatterers from different ovals.

2.6 SPATIAL STRUCTURE OF MULTIPLE ANTENNA CHANNELS

As previously mentioned, one of the benefits of multiple antenna systems is their ability to resolve the channel spatially. That is, we may infer the location of scatterers in the channel using a multiple antenna system. This section contains a brief discussion of *beamformers* and the *MIMO azimuth power spectrum* (APS). The MIMO APS is a useful tool in visualizing the spatial structure of the MIMO channel. In Section 3.4.3, we use the MIMO APS to show how the Kronecker model distorts the spatial structure of a given channel. In Chapter 6, we use beamformers to evaluate the structured model's ability to approximate the spatial structure of a given channel and compare it with results from the Kronecker model.

2.6.1 SIMO Channels and Beamformers

By way of explanation, we begin with a discussion of *plane waves*, and how one can infer the incident angle of a plane wave using an antenna array. We then discuss beamformers as a method of visualizing incoming plane waves from different directions.

2.6.1.1 Plane Waves and Spatial Structure   Real-life radios work by radiating energy into a medium using an antenna. Antennas can be *directive* in that they radiate the majority of their energy in one direction or *omnidirectional* such that they radiate energy in many directions at once. The *antenna pattern* is a plot of the energy radiated by a given antenna as a function of angle in all directions. An *infinitesimal point source* is an omnidirectional antenna that radiates energy equally in all directions and thus has a spherical antenna pattern. This type of antenna is impossible to realize in practice but serves well to illustrate the theory.

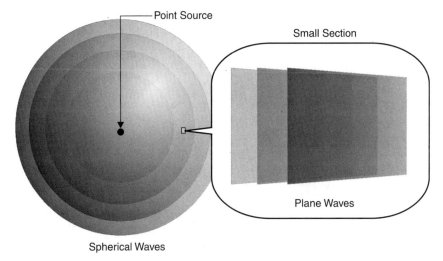

Figure 2.11 Visualization of a plane wave, showing how a spherical antenna pattern, at a distance, can be well approximated as a plane wave.

Consider the spherical wave pattern shown in Fig. 2.11. If the receiver is separated from the transmitter by a considerable distance, the receive-antenna captures only a small cross section of the energy radiated by the transmitter. The effective cross section is referred to as the *antenna aperture*. At small distances, because of the curvature of the sphere, the energy captured by the receive-antenna can vary across its aperture. As the distance between the transmitter and receiver increases, the sphere impinging each receive-antenna gets bigger, and the energy can be considered to be constant across a given receive-antenna's aperture. In this case, the sphere cross section can be approximated as a flat plane, or a *plane wave*.

At some point, as the distance between the transmitter and the receiver increases, the radiated energy from any antenna can be approximated as a plane wave, regardless of its radiation pattern.

Consider the case illustrated in Fig. 2.12, where there are three distinct paths from the transmitter to the receiver. In addition to the line-of-sight plane wave, two other plane waves consisting of the reflected energy from both scatterers impinge on the receive array. The incident directions of all three plane waves can be inferred using the information from all receive antennas. In real-life channels, plane waves are often reflected from single scatterers or groups of scatterers. From the receiver's point of view, the physical channel is simply a collection of plane waves, impinging from many possible directions. By resolving the direction of a plane wave or groups of plane waves, we can resolve the direction of a scatterer or groups of scatterers relative to the receive array. This implies that, with multiple antenna systems, the *spatial structure* of the channel, or the orientation of scatterers in the channel, can be resolved. A given system's ability to resolve plane waves, or its *spatial resolution*, depends on a number of factors, most notably the number of antennas, their individual patterns, and the array configuration.

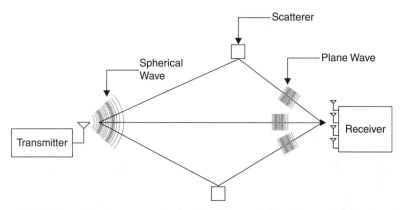

Figure 2.12 Transmitted energy reflects off scatterers in the channel and arrives at the receiver as a group of plane waves, with each impinging from a different direction.

2.6.1.2 Steering Vector

Here, we consider an example 2×1 SIMO system. Let the two receive-antennas be separated by distance d_{ant}, as shown in Fig. 2.13.

A plane wave is incident on the array at angle φ with respect to the array azimuth. As φ increases from $0°$ (the plane wave impinges on the array at *boresight*) to $90°$ (the plane wave impinges edgewise onto the array), the plane wave takes more time to cover the distance $d \sin \varphi$, and thus it is possible to distinguish from what angle the plane wave was incident. Statistically speaking, the signals from both antennas will correlate in the direction of incidence, as the same signal is impinging on both antennas.

The array response in a given direction is quantified by the so-called *steering vector*. By way of explanation, we begin with the most general case. Consider an arbitrary receive array of M_{Rx} antennas. Given an arbitrary origin, the location of each antenna element is defined by the vector $\mathbf{r}_m$ for $1 \leq m \leq M_{Rx}$. The mth antenna pattern is defined in 3-space by the function $f_m(\varphi, \theta)$, where φ and θ are the azimuth and elevation angles, respectively. This is illustrated in Fig. 2.14. The ℓth plane wave undergoes a phase shift of $e^{j\frac{2\pi}{\lambda} r_m^T \cos \varphi_\ell \sin \vartheta_\ell}$ by the time it reaches antenna m. The

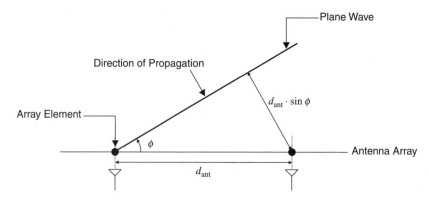

Figure 2.13 Plane wave impinging on two linear array elements.

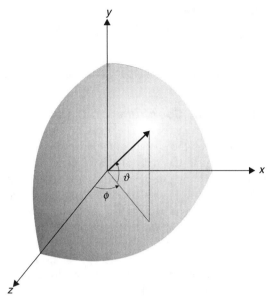

Figure 2.14 The azimuth and elevation angles φ and θ, with respect to an antenna element.

response of the array to this plane wave is weighted by the antenna pattern in that direction. Thus, the steering vector for an arbitrary array for the ℓth plane wave is defined as [65, 66]

$$
\mathbf{a}_\ell(\varphi,\,\vartheta) = \alpha_\ell
\begin{bmatrix}
e^{j\frac{2\pi}{\lambda} r_1^T \cos \varphi_\ell \sin \vartheta_\ell} \cdot f_1(\varphi,\,\vartheta) \\
e^{j\frac{2\pi}{\lambda} r_2^T \cos \varphi_\ell \sin \vartheta_\ell} \cdot f_2(\varphi,\,\vartheta) \\
\vdots \\
e^{j\frac{2\pi}{\lambda} r_{M_{\mathrm{Rx}}}^T \cos \varphi_\ell \sin \vartheta_\ell} \cdot f_{M_{\mathrm{Rx}}}(\varphi,\,\vartheta)
\end{bmatrix},
\tag{2.49}
$$

where α_ℓ is the complex amplitude of the ℓth plane wave.

Often, we speak of a *uniform linear array* (ULA) in which each antenna is said to be *isotropic*, that is $f_m(\varphi, \theta) = 1$ for all m, and each antenna is separated by distance d_{ant} from its neighbor. We can also use the first antenna as a phase/location reference, and we only consider the array response in the azimuth. In this case, the steering vector depends only on φ, and (2.49) reduces to the Fourier vector

$$
\mathbf{a}_\ell(\varphi) = \alpha_\ell
\begin{bmatrix}
1 \\
e^{j2\pi 2\frac{\delta}{\lambda} \sin \varphi_\ell} \\
\vdots \\
e^{j2\pi M_{\mathrm{Rx}}\frac{\delta}{\lambda} \sin \varphi_\ell}
\end{bmatrix}.
\tag{2.50}
$$

There are instances where the energy impinging on an array is not planar. These include the case where there are scatterers close to the array (versus the array size) and when we have diffuse scattering. In this case, the steering vector depends on the geometry of the incident wave, and (2.50) does not hold. In the following, we assume that the majority of the received energy is in the form of plane waves.

2.6.1.3 Azimuth Power Spectrum and Vector Beamformers

A fundamental assertion of correlative channel models is that the correlation between paths completely characterizes the spatial structure of the channel. The following discusses how we can estimate the APS of the vector channel given the channel correlation, which *varies* according to the channel, and array steering vector, which remains *fixed*. The APS allows us to visualize the location of scatterers in the azimuth plane. We estimate the APS using a beamformer. In general, there are many different beamformers. Here, we consider the *Bartlett beamformer* [67], one of the simplest beamformers.

First we wish to make the connection between the average received power and $\mathbf{R}_h$. Consider the noiseless narrowband MISO vector channel $\mathbf{h}_{\mathrm{MISO}}$, for which the input–output relationship through a noiseless channel can be written as

$$y(t) = \mathbf{h}_{\mathrm{MISO}}^{\mathrm{T}} \mathbf{x}(t). \tag{2.51}$$

If we consider only a single time instant, we can drop the dependence on t. The relationship can be recast as

$$
\begin{aligned}
y &= \mathbf{h}_{\mathrm{MISO}}^{\mathrm{T}} \mathbf{x} \\
&= \mathbf{x}^{\mathrm{T}} \mathbf{h}_{\mathrm{MISO}}. \tag{2.52}
\end{aligned}
$$

The mean (average) received power is computed as the expected value of the complex inner-product; that is,

$$
\begin{aligned}
\mathrm{E}\big\{|y|^2\big\} &= \mathrm{E}\{yy^*\} \\
&= \mathrm{E}\{\mathbf{x}^T \mathbf{h}_{\mathrm{MISO}} \mathbf{h}_{\mathrm{MISO}}^H \mathbf{x}^*\}. \tag{2.53}
\end{aligned}
$$

Most often, to focus on the statistics of the channel, the transmit vector $\mathbf{x}$ is assumed to be known or fixed. In this case, we may move the expectation inward,

$$
\begin{aligned}
\mathrm{E}\{yy^*\} &= \mathbf{x}^T \mathrm{E}\{\mathbf{h}_{\mathrm{MISO}} \mathbf{h}_{\mathrm{MISO}}^H\} \mathbf{x}^* \\
&= \mathbf{x}^T \mathbf{R}_h \mathbf{x}^*, \tag{2.54}
\end{aligned}
$$

where $\mathrm{E}\{\mathbf{h}_{\mathrm{MISO}} \mathbf{h}_{\mathrm{MISO}}^H\} = \mathbf{R}_h$ by definition. Thus, we can compute the average received power using $\mathbf{R}_h$. We make use of this fact in the following.

Beamformers work by estimating the incident power at a specific azimuth angle. To do this, they require some *spatial filter* $\mathbf{f}_{\mathrm{space}}(\varphi)$ that focuses on incident energy from a single direction. Let $\hat{P}(\varphi)$ denote the average power incident from φ. Using $\mathbf{f}_{\text{space}}(\varphi)$, we can rewrite (2.54) to estimate $\hat{P}(\varphi)$ as

$$\hat{P}(\varphi) = \mathbf{f}_{\text{space}}^{T}(\varphi)\mathbf{R}_{h}\mathbf{f}_{\text{space}}^{*}(\varphi). \tag{2.55}$$

In the case of a Bartlett beamformer, the optimal spatial filter for a uniform linear array is $\mathbf{a}(\varphi)$, where

$$\mathbf{a}(\varphi) = \begin{bmatrix} 1 \cdot f_1(\varphi) \\ e^{j2\pi 2\frac{\delta}{\lambda}\sin\varphi}f_2(\varphi) \\ \vdots \\ e^{j2\pi M_{\text{Rx}}\frac{\delta}{\lambda}\sin\varphi}f_{M_{\text{Rx}}}(\varphi) \end{bmatrix}, \tag{2.56}$$

and $f_m(\varphi)$ is the azimuth antenna pattern for antenna m, and M_{Rx} is the number of array elements. A Bartlett beamformer thus assumes the form

$$f_{\text{space,Bart}}(\varphi) = \frac{\mathbf{a}^{*}(\varphi)}{\sqrt{\mathbf{a}^{H}(\varphi)\mathbf{a}(\varphi)}}, \tag{2.57}$$

where $f_{\text{space,Bart}}(\varphi)$ is normalized such that $\left\| f_{\text{space,Bart}}(\varphi) \right\|_{2} = 1$. Substituting $f_{\text{space,Bart}}(\varphi)$ into (2.55), we get the *Bartlett spectrum* $\hat{P}_{\text{Bart}}(\varphi)$, defined by

$$\hat{P}_{\text{Bart}}(\varphi) = \frac{\mathbf{a}^{H}(\varphi)\mathbf{R}_{h}\mathbf{a}(\varphi)}{\mathbf{a}^{H}(\varphi)\mathbf{a}(\varphi)}. \tag{2.58}$$

In the special case where we have a ULA, $\mathbf{a}(\varphi)$ takes on the form of the Fourier vector (2.50), and the Bartlett spectrum becomes the Fourier transform of $\mathbf{R}_{h}$. It is important to note that, for a given array, $\mathbf{a}(\varphi)$ is assumed to be known, and $\mathbf{R}_{h}$ depends on the composition of the physical channel. Thus, the vector correlation matrix completely characterizes the spatial structure of the vector channel.

2.6.2 MIMO Beamformers

The concept of a beamformer can be extended to the MIMO channel. For the vector channel, the vector beamformer gave a picture of the average energy in a specific azimuth direction. For the MIMO channel, a MIMO beamformer estimates the average power coupled between a given transmit-angle and receive-angle.

We refer to the transmit-angle φ_{Tx} as the *angle of departure* (AoD) and receive-angle φ_{Rx} as the *angle of arrival* (AoA). Given the *transmit steering vector* $\mathbf{a}_{\text{Tx}}(\varphi_{\text{Tx}})$ and the *receive steering vector* $\mathbf{a}_{\text{Rx}}(\varphi_{\text{Rx}})$, the *MIMO Bartlett spectrum* is defined as

$$\hat{P}_{\text{Bart}}(\varphi_{\text{Tx}}, \varphi_{\text{Rx}}) = [\mathbf{a}_{\text{Tx}}(\varphi_{\text{Tx}}) \otimes \mathbf{a}_{\text{Rx}}(\varphi_{\text{Rx}})]^{H}\mathbf{R}_{H}[\mathbf{a}_{\text{Tx}}(\varphi_{\text{Tx}}) \otimes \mathbf{a}_{\text{Rx}}(\varphi_{\text{Rx}})], \tag{2.59}$$

where $\otimes$ denotes the Kronecker product, and $\mathbf{R}_{H}$ is the narrowband correlation matrix.

Note that $P_{Bart}(\varphi_{Tx}, \varphi_{Rx})$ is necessarily a two-dimensional function. In the literature, $P_{Bart}(\varphi_{Tx}, \varphi_{Rx})$ is often plotted as a color map on a decibel (dB) scale. Let $APS_{MIMO}(\varphi_{Rx}, \varphi_{Tx})$ be the *MIMO APS*. For the Bartlett beamformer, the MIMO APS is computed as

$$APS_{MIMO}(\varphi_{Rx}, \varphi_{Tx}) = 10 \cdot \log_{10}[P_{Bart}(\varphi_{Tx}, \varphi_{Rx})] \text{ dB.} \tag{2.60}$$

Peaks in the spectrum represent a resolvable scatterer or a cluster of scatterers. We refer to these as *clusters*. Each cluster represents a resolvable *path* from the transmitter to the receiver. The properties of a MIMO channel are largely determined by the spatial structure of the channel; the number and orientation of the scatterers determines the degree of spatial diversity and the multiplexing gain of the channel. Thus, the MIMO APS can be used to qualitatively evaluate a given channel model's ability to approximate the spatial structure of the channel. As with the vector case, given that $\mathbf{a}_{Tx}(\varphi_{Tx})$, $\mathbf{a}_{Rx}(\varphi_{Rx})$ are fixed quantities, $\mathbf{R}_H$ completely characterizes the spatial structure of the matrix channel.

MIMO Beamformer Example   We now wish to illustrate the utility of a MIMO beamformer with respect to the EVD of the narrowband correlation matrix. In the following, we show how to generate a synthetic narrowband MIMO channel consisting of S resolvable scatterers. Let $(\varphi_{Rx,s}, \varphi_{Tx,s})$ denote the position of the sth scatterer on the AoA–AoD plane. The number and behavior of scatterers in the channel can be determined by using the EVD to decompose $\mathbf{R}_H$. In the following example, we summarize results from Ref. 51.

Consider the case where $\mathbf{R}_H$ has only one significant eigenvalue. Furthermore, let there be S correlated scatterers. The location of the sth scatterer is determined by the sth impinging wave. In AoA–AoD space, this is described by the outer product of the vectors

$$\mathbf{a}_s(\varphi_{Rx,s}, \varphi_{Tx,s}) = \mathbf{a}_{Rx}(\varphi_{Rx,s})\mathbf{a}_{Tx}^T(\varphi_{Tx,s}). \tag{2.61}$$

Assume that the transmitter and receiver are equipped with ULAs. In this case, $\mathbf{a}_{Rx}(\varphi_{Rx,s})$ and $\mathbf{a}_{Tx}(\varphi_{Tx,s})$ are given by

$$\mathbf{a}_{Rx}(\varphi_{Rx,s}) = \begin{bmatrix} 1 \\ e^{j2\pi 2\frac{\delta}{\lambda}\sin\varphi_{Rx,s}} \\ \vdots \\ e^{j2\pi M_{Rx}\frac{\delta}{\lambda}\sin\varphi_{Rx,s}} \end{bmatrix}$$

$$\mathbf{a}_{Tx}(\varphi_{Tx,s}) = \begin{bmatrix} 1 \\ e^{j2\pi 2\frac{\delta}{\lambda}\sin\varphi_{Tx,s}} \\ \vdots \\ e^{j2\pi M_{Rx}\frac{\delta}{\lambda}\sin\varphi_{Tx,s}} \end{bmatrix}. \tag{2.62}$$

The first eigenbasis $\mathbf{U}_1$ of $\mathbf{R}_H$ is defined as

$$\mathbf{U}_1 = \sum_{s=1}^{S} \mathbf{a}_s(\varphi_{\text{Rx},s}, \varphi_{\text{Tx},s})$$

$$= \sum_{s=1}^{S} \mathbf{a}_{\text{Rx}}(\varphi_{\text{Rx},s}) \mathbf{a}_{\text{Tx}}^T(\varphi_{\text{Tx},s}), \tag{2.63}$$

where the pair $(\varphi_{\text{Rx},s}, \varphi_{\text{Tx},s})$ specifies the AoA and AoD of the sth scatterer, respectively, and we let $\lambda_1 = 1$. Let all other eigenvalues of $\mathbf{R}_H$ be zero; that is, $\lambda_k = 0$ for $k = 2, \ldots, M_{\text{Rx}}M_{\text{Tx}}$. Then, the sole eigenvector of $\mathbf{R}_H$ given $\mathbf{U}_1$ is computed as

$$\mathbf{u}_1 = \text{vec}(\mathbf{U}_1). \tag{2.64}$$

$\mathbf{R}_H$ is computed using the EVD as

$$\mathbf{R}_H = \sum_{k=1}^{M_{\text{Rx}}M_{\text{Tx}}} \mathbf{u}_k \mathbf{u}_k^H$$

$$= \mathbf{u}_1 \mathbf{u}_1^H. \tag{2.65}$$

Figure 2.15 illustrates the exemplar MIMO APS using an 8×8 system equipped with ULAs, whose antenna elements are separated by $d_{\text{ant}} = \lambda/2$. The channel consists of $S = 4$ equally weighted and correlated scatterers given by $(\varphi_{\text{Rx},s}, \varphi_{\text{Tx},s}) = \{(0°, -50°), (0°, 0°), (60°, 40°), (-50°, 60°)\}$.

In general, the energy in scatterers closer to the edges of the MIMO APS tend to "smear" toward the edge. Also, unless the scatterers associated with each eigenbasis

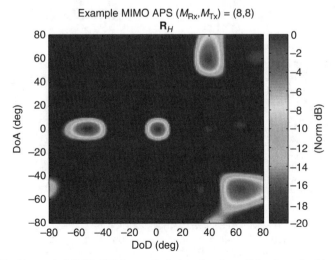

Figure 2.15 Example MIMO APS for an 8×8 showing four distinct resolvable scatterers.

of $\mathbf{R}_H$ are plotted individually, there is no way to distinguish between correlated and uncorrelated scatterers using the MIMO APS alone.

Now consider the case where there are four uncorrelated scatterers. Let the location of each scatterer be $(\varphi_{\text{Rx},s}, \varphi_{\text{Tx},s}) = \{(0°, -50°), (0°, 0°), (60°, 40°), (-50°, 60°)\}$. In this case, $\mathbf{R}_H$ consists of four eigenbases

$$\mathbf{U}_k = \mathbf{a}_k(\varphi_{\text{Rx},k}, \varphi_{\text{Tx},k})$$
$$= \mathbf{a}_{\text{Rx}}(\varphi_{\text{Rx},k})\mathbf{a}_{\text{Tx}}^T(\varphi_{\text{Tx},k}) \quad (2.66)$$

for $k = \{1, \ldots, 4\}$. The MIMO APS of $\mathbf{R}_H$ associated with a single eigenvalue is computed by first computing the matrix

$$\mathbf{R}_{H,k} = \lambda_k \mathbf{u}_k \mathbf{u}_k^H \quad (2.67)$$

for $k = \{1, \ldots, 4\}$. The Bartlett beamformer for each $\mathbf{R}_{H,k}$ is computed as

$$P_{\text{Bart},k}(\varphi_{\text{Tx},k}, \varphi_{\text{Rx},k}) = (\mathbf{a}_{\text{Tx}}(\varphi_{\text{Tx},k}) \otimes \mathbf{a}_{\text{Rx}}(\varphi_{\text{Rx},k}))^H$$
$$\times \mathbf{R}_{H,k}(\mathbf{a}_{\text{Tx}}(\varphi_{\text{Tx},k}) \otimes \mathbf{a}_{\text{Rx}}(\varphi_{\text{Rx},k})), \quad (2.68)$$

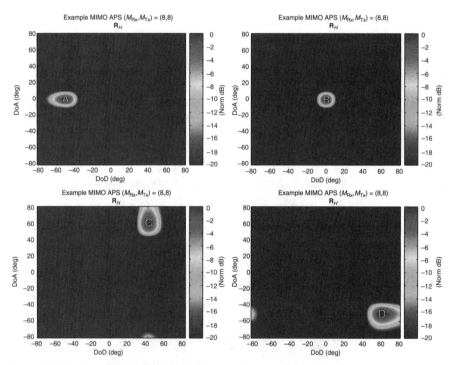

Figure 2.16   Example MIMO APS showing the scatterer described by each eigenvalue. Note that the complete MIMO APS, formed by summing the contribution from all eigenvalues, resembles the MIMO APS shown in Fig. 2.15, which was formed using four *correlated* scatterers.

where $\otimes$ is the Kronecker product, and the MIMO APS for each eigenvalues as

$$\text{APS}_{\text{MIMO},k}(\varphi_{\text{Rx}}, \varphi_{\text{Tx}}) = 10 \cdot \log_{10}(P_{\text{Bart},k}(\varphi_{\text{Tx}}, \varphi_{\text{Rx}})). \tag{2.69}$$

Let all four eigenvalues $\lambda_k = 1$. This means that all four scatter clusters are equally weighted. The MIMO APS associated with all four eigenvalues is plotted in Fig. 2.16. The MIMO APS of $\mathbf{R}_H$ is computed by first reconstituting it from all $\mathbf{R}_{H,k}$, that is,

$$\mathbf{R}_H = \sum_{k=1}^{4} \mathbf{R}_{H,k}, \tag{2.70}$$

and computing the MIMO APS for $\mathbf{R}_H$. If we do this, we get the *same MIMO APS* of Fig. 2.15. This means that the MIMO APS alone does not give us enough information about the correlation between scatterers in a channel. We need to plot the MIMO APS associated with each eigenvalue of $\mathbf{R}_H$ to determine the statistical relationship between scatterers in the channel.

2.7 SUMMARY AND DISCUSSION

2.7.1 Channel Classifications

In this chapter, we defined concepts used throughout the remainder of the book. The analytical channel is a conceptualization of the physical channel. It characterizes the channel as a linear filter. The channel can be classified as narrowband or wideband and time-variant or time-invariant. The discussion in the remainder of the book covers the special cases of narrowband and wideband time-invariant channels.

Bello's model is an elegant way of describing the input–output relationships for a time-variant wideband channel. Bello's model uses the Fourier transform to character-ize both the time-variant and time-dispersive nature of the channel in the time and Fourier domain. Bello's model led to several concepts important to the characteriz-ation of the wideband channel, such as the tapped delay-line filter and the RMS delay spread metric. The RMS delay spread is used to quantify a channel's time-dispersion and is defined in the context of cluster models in Chapter 4. The point at which a channel is considered to be time-variant or time-dispersive depends on the application. Time-variation and time-dispersion can be quantified by their Doppler and RMS delay spread, respectively, but there is no universally agreed upon point past which a channel is considered to be time-variant or wideband, respectively.

2.7.2 Multiantenna Channels

We then progress to multiantenna channels. Diversity is used to combat the effects of multipath fading, where Rayleigh fading is a special case. Diversity systems employ either multiple antennas at the receiver or the transmitter. MIMO systems are a natural extension of diversity systems and are equipped with multiple antennas at the receiver and the transmitter. The key difference between diversity and MIMO systems is that the latter can trade off diversity for increased capacity. This represents an added degree of freedom in system design: Diversity increases reliability in bad channels, and multiplexing gain can be used to increase spectral efficiencies where bandwidth is a premium.

After defining the MIMO channel, we recast the discrete tapped delay-line model using tensor calculus. The tensor system model is an elegant way of representing the wideband MIMO channel and allows the use of tensor decomposition. The H-tensor $\mathcal{H}$ is an elegant way to express the complex gains in a wideband MIMO channel. Representing the channel as a third-order tensor implies that we can use tensor decomposition to derive a new channel model. This has the advantage of maintaining more of the channel structure. We use tensor calculus to compute the sixth-order wideband correlation tensor $\mathcal{R}$. This representation is unique in that it describes correlation across transmit-receive-delay space, which would be analytically difficult otherwise. The wideband correlation matrix $\mathbf{R}_{\mathrm{WB},H}$ is one possible mapping of the elements of $\mathcal{R}$ to a matrix. The structured model, presented in Section 3.6, is derived using the tensor decomposition of the wideband MIMO channel.

2.7.3 Spatial Structure and the APS

With multiantenna systems, we gain the ability to resolve the AoA and AoD of plane waves at the receiver and transmitter, respectively. The APS is a useful tool in visualizing the position and density of scatterers in a channel. The first step in determining the APS is to compute the second-order statistics of the channel. In Section 2.4, we began by defining the second-order statistics of the narrowband vector channel. This progressed naturally into the narrowband and wideband MIMO channels. In addition, the EVD of the channel correlation allows us to classify the scatterers in the channel as correlated and uncorrelated and brings structure to the channel. In Section 2.6, we defined the spatial structure as the orientation of scatterers in a channel and showed how the APS can be used to visualize the spatial structure. The APS is used in Chapter 6 as a qualitative metric to compare the performance of the structured versus Kronecker models. The APS, extended to include the elevation and delay domains, also forms a cornerstone for cluster modeling. Chapter 4 covers the fundamentals of cluster models.

2.8 NOTES AND REFERENCES

2.8.1 Channel Classifications

Linear time-invariant system theory is a mature topic and is the subject of many textbooks, including Ref. 68. Time-invariant narrowband channels are a special case of the linear time-invariant channel.

The study of linear time-varying filters in a system context dates back to the work of Zadeh (69), Kailath (70), and Bello (40). Of the three, Bello is often credited with the most complete characterization of the time-varying, time-dispersive radio channel. The tapped delay-line model is a direct consequence of Bello's model.

Jakes (12) provides one of the most complete introductions to diversity channels. Work in diversity channels led naturally to the founding work of Winters (15), Foschini (16), and Telatar (17) in MIMO channel theory.

Durgin (60) presents a more complete treatment of Bello's model and extends the concepts to higher-order domains.

2.8.2 Second-Order Statistics of Multiantenna Channels

Jakes (12) presents an in-depth analysis of correlation in narrowband diversity channels. Numerous ray-tracing and scattering models that came after it were based on Jakes' model.

D.-S. Shiu et al. (31) and Chizhik et al. (35) analyze the effect of correlation between paths to MIMO capacity.

In his dissertation (51), Weichselberger provides one of the most complete analyses of the EVD of the correlation in narrowband MIMO channels and relates this to the statistical relationship between scatterers.

2.8.3 The Spatial Structure of Multiantenna Channels

Diversity beamformers and the MIMO APS are discussed in detail in Ref. 51, including the relationship of the MIMO APS to the EVD of $\mathbf{R}_H$.

A comprehensive review of digital beamforming techniques can be found in Ref. 66.

3

CORRELATIVE MODELS

This chapter presents a review of several correlative channel models; namely, the *Kronecker*, *Weichselberger*, and *structured* models. Due to its simplicity, the Kronecker model is by far the most popular MIMO channel model in the literature. It greatly simplifies channel analysis, as it holds that scatterers around the transmitter fade independently of those around the receiver. Despite its popularity, the Kronecker model has been shown to be inaccurate. The Weichselberger model uses the eigenvalue decomposition (EVD) of the channel as its parameters, and, in most cases, is more accurate than the Kronecker model. The structured model is an extension of the Weichselberger model to the wideband MIMO channel. It uses tensors to express the wideband MIMO channel in an elegant fashion, often with fewer parameters than the Kronecker model. In Chapter 6, we show that the structured model consistently outperforms the Kronecker model.

The Kronecker model is capable of modeling the narrowband and wideband MIMO channels, whereas the Weichselberger model can be applied to narrowband MIMO channels only. The structured model describes the wideband MIMO channel and reduces to the Weichselberger model in the narrowband case.

We begin by discussing synthesis equations for the vector and matrix channels. These synthesis equations use the full correlation to generate an ensemble of channels with the same spatial characteristics as a given channel. Because of this, they require the most parameters. To reduce the number of parameters, the Kronecker, Weichselberger, and structured models use the concept of *one-sided correlation* in

Multiple-Input, Multiple-Output Channel Models: Theory and Practice. By Nelson Costa and Simon Haykin
Copyright © 2010 John Wiley & Sons, Inc.

deriving their models. Before summarizing the Kronecker and Weichselberger models, we discuss one-sided correlation in narrowband MIMO channels. In Section 3.6.2, we use tensor calculus to extend the idea of one-sided correlation to include wideband MIMO channels. This leads naturally to the derivation of the structured model.

3.1 VECTOR CHANNEL SYNTHESIS FROM THE VECTOR CORRELATION MATRIX

For a complex-Gaussian distributed vector channel, the synthesis equation is [51]

$$\mathbf{h}_{\text{synth}} = \mathbf{R}_h^{1/2}\mathbf{g}_h, \tag{3.1}$$

where $\mathbf{h}_{\text{synth}}$ is the *synthetic h-vector*, and $\mathbf{R}_h^{1/2}$ is the square-root matrix of $\mathbf{R}_h$, which is defined below. The elements $\mathbf{g}_h \in \mathbb{C}^{M \times 1}$ are *independent and identically distributed* (IID) complex-Gaussian random variables, with zero mean and unit variance. The vector $\mathbf{g}_h$ is normalized such that $\mathrm{E}\{\mathbf{g}_h\mathbf{g}_h^H\} = \mathbf{I}$. The implicit model assumption is that the complex gains of the original channel are complex-Gaussian distributed random variables. The Gaussian assumption is popular in the literature and has been shown experimentally to model the statistics of many physical channels (see, for example, Ref. 12).

Because $\mathbf{R}_h$ is Hermitian symmetric and positive semidefinite, we can compute its square root via the EVD. That is, given

$$\mathbf{R}_h = \mathbf{U}_h\mathbf{D}_h\mathbf{U}_h^H, \tag{3.2}$$

the square-root matrix is defined by

$$\mathbf{R}_h^{1/2} = \mathbf{U}_h\mathbf{D}_h^{1/2}\mathbf{U}_h^H = \sum_{k=1}^{M}\lambda_k^{1/2}\mathbf{u}_k\mathbf{u}_k^H, \tag{3.3}$$

where

$$\mathbf{D}_h^{1/2} = \begin{pmatrix} \lambda_1^{1/2} & 0 & \cdots & 0 \\ 0 & \lambda_2^{1/2} & \cdots & 0 \\ \vdots & \vdots & \ddots & \vdots \\ 0 & 0 & \cdots & \lambda_M^{1/2} \end{pmatrix}, \tag{3.4}$$

and λ_m are the eigenvalues of $\mathbf{R}_h$ for $m \in \{1, \ldots, M\}$. Note that the square-root matrix satisfies the condition $\mathbf{R}_h^{1/2}\mathbf{R}_h^{H/2} = \mathbf{R}_h$.

Analytically, we can check whether $\mathbf{h}_{\text{synth}}$ has the same spatial characteristics as $\mathbf{R}_h$. To do this, we compute the correlation matrix of $\mathbf{h}_{\text{synth}}$ as

$$
\begin{aligned}
\mathrm{E}\{\mathbf{h}_{\text{synth}}\mathbf{h}_{\text{synth}}^H\} &= \mathrm{E}\left\{\mathbf{R}_h^{1/2}\mathbf{g}_h\mathbf{g}_h^H\mathbf{R}_h^{H/2}\right\} \\
&= \mathbf{R}_h^{1/2}\mathrm{E}\{\mathbf{g}_h\mathbf{g}_h^H\}\mathbf{R}_h^{H/2} \\
&= \mathbf{R}_h.
\end{aligned}
\tag{3.5}
$$

This implies that the correlation matrix of $\mathbf{h}_{\text{synth}}$ and $\mathbf{h}$ are identical, and thus the spatial properties are preserved by the model.

Using the EVD of $\mathbf{R}_h$, we can rewrite (3.1) as

$$
\mathbf{h}_{\text{synth}} = \sum_{k=1}^{M} g_k \lambda_{h,k}^{1/2} \mathbf{u}_{h,k},
\tag{3.6}
$$

where g_k are the elements of $\mathbf{g}_h$, and the eigenvectors $\mathbf{u}_{h,k}$ are the columns of the eigen-basis $\mathbf{U}_h$; that is, $\mathbf{U}_h = [\mathbf{u}_{h,1} \ \mathbf{u}_{h,1} \ \cdots \ \mathbf{u}_{h,M}]$. Note that (3.1) is equivalent to (3.6) via the EVD of $\mathbf{R}_h$.

3.2 MATRIX CHANNEL SYNTHESIS FROM THE NARROWBAND CORRELATION MATRIX

By extension, (3.1) can be applied to the narrowband MIMO channel using the unvec($\cdot$) operator, viz. (51),

$$
\mathbf{H}_{\text{synth}} = \mathrm{unvec}(\mathbf{R}_H^{1/2}\mathbf{g}_H),
\tag{3.7}
$$

where $\mathbf{g}_H \in \mathbb{C}^{M_{\text{Rx}}M_{\text{Tx}} \times 1}$ is a complex-Gaussian IID vector, and $\mathbf{R}_H^{1/2}$ is the square-root matrix of $\mathbf{R}_H$. We refer to $\mathbf{H}_{\text{synth}}$ as the *synthetic H-matrix*.

3.2.1 Number of Model Parameters

For many channel models, it is important to highlight the number of *parameters*. The parameters of a channel model are any variables that depend on the physical channel. In general, increasing the number of model parameters also increases its complexity. Thus, it is important to keep the number of parameters in the model as low as possible while still modeling the channel reasonably accurately. The number of parameters required to model the channel can be reduced by imposing some structure.

Equation (3.7) involves an $M_{\text{Rx}}M_{\text{Tx}} \times 1$ vector $\mathbf{g}_H$ and the $M_{\text{Rx}}M_{\text{Tx}} \times M_{\text{Rx}}M_{\text{Tx}}$ matrix $\mathbf{R}_H$. The elements in $\mathbf{g}$ are generated randomly; these are not included as parameters in the model. However, $\mathbf{R}_H$ must be included for every new channel considered. Thus, (3.7) involves $(M_{\text{Rx}}M_{\text{Tx}})^2$ parameters.

As with the vector case, we recast (3.7) using the EVD of $\mathbf{R}_H^{1/2}$ as

$$
\mathbf{H}_{\text{synth}} = \sum_{k=1}^{M_{\text{Rx}}M_{\text{Tx}}} g_k \lambda_k^{1/2} \mathbf{U}_k,
\tag{3.8}
$$

where g_k are complex-Gaussian random variables, λ_k is the kth eigenvalues of $\mathbf{R}_H$, and $\mathbf{U}_k$ is the kth eigenbasis of $\mathbf{R}_H$. We note that (3.8) is equivalent to (3.7) via the EVD of $\mathbf{R}_H$. However, because of the $M_{\mathrm{Rx}} \times M_{\mathrm{Tx}}$ eigenvalues λ_k, (3.8) requires more parameters to compute $\mathbf{H}_{\mathrm{synth}}$ than does (3.7). Specifically, (3.8) requires $(M_{\mathrm{Rx}}M_{\mathrm{Tx}})^2 \times M_{\mathrm{Rx}}M_{\mathrm{Tx}}$ parameters versus $(M_{\mathrm{Rx}}M_{\mathrm{Tx}})^2$ for (3.7). However, the latter equation imposes more structure than the former; (3.8) breaks the channel into $M_{\mathrm{Rx}} \cdot M_{\mathrm{Tx}}$ parallel channels, each with power λ_k.

3.3 ONE-SIDED CORRELATION FOR NARROWBAND MIMO CHANNELS

In previous sections, we covered ways in which we can generate exemplar H-matrices and H-tensors using the full correlation matrix. The number of parameters needed to synthesize new channels can be reduced by somehow approximating the full correlation. One technique for doing this involves decomposing the full correlation using the concept of *one-sided correlation*. The following summarizes one-sided correlation and its relation to $\mathbf{R}_H$ (51).

Consider the narrowband channel gain matrix $\mathbf{H}$, whose structure we repeat here for convenience of presentation:

$$\mathbf{H} = \begin{pmatrix} h_{11} & \cdots & h_{1M_{\mathrm{Tx}}} \\ \vdots & \ddots & \vdots \\ h_{M_{\mathrm{Rx}}1} & \cdots & h_{M_{\mathrm{Rx}}M_{\mathrm{Tx}}} \end{pmatrix}. \tag{3.9}$$

The row vector $\mathbf{h}_{\mathrm{row},m}$ is defined as the mth row and $\mathbf{h}_{\mathrm{col},n}$ as the nth column of $\mathbf{H}$, viz.,

$$\mathbf{h}_{\mathrm{row},m} = [h_{m1} \quad h_{m2} \quad \cdots \quad h_{mM_{\mathrm{Tx}}}], \tag{3.10}$$

and

$$\mathbf{h}_{\mathrm{col},n} = [h_{1n} \quad h_{2n} \quad \cdots \quad h_{M_{\mathrm{Rx}}n}]^T. \tag{3.11}$$

The *narrowband receive-correlation matrix* $\mathbf{R}_{\mathrm{Rx}}$ is defined as

$$\mathbf{R}_{\mathrm{Rx}} = \mathrm{E}\{\mathbf{H}\mathbf{H}^H\}$$
$$= \sum_{n=1}^{M_{\mathrm{Tx}}} \mathrm{E}\{\mathbf{h}_{\mathrm{col},n}\mathbf{h}_{\mathrm{col},n}^H\}, \tag{3.12}$$

and the *narrowband transmit-correlation matrix* $\mathbf{R}_{\mathrm{Tx}}$ as

$$\mathbf{R}_{\mathrm{Tx}} = \mathrm{E}\{\mathbf{H}^T\mathbf{H}^*\}$$
$$= \sum_{m=1}^{M_{\mathrm{Rx}}} \mathrm{E}\{\mathbf{h}_{\mathrm{row},m}^T\mathbf{h}_{\mathrm{row},m}^*\}. \tag{3.13}$$

Note that $\mathbf{h}_{\text{row},m}$ is the MISO channel formed by all transmitters to receiver m. It can be thought of as the vector channel formed by considering all elements h_{mn} with m fixed, and $n = \{1, \ldots, M_{\text{Tx}}\}$. Similarly, $\mathbf{h}_{\text{col},n}$ is the SIMO channel formed by transmitter n to all the receivers. It is the vector channel formed by considering all elements h_{mn} with n fixed and $m = \{1, \ldots, M_{\text{Rx}}\}$. In this way, the receive-correlation matrix $\mathbf{R}_{\text{Rx}}$ can be viewed as the correlation across all receivers for a fixed transmitter. Conversely, the transmit-correlation matrix $\mathbf{R}_{\text{Tx}}$ can be interpreted as the correlation across all transmitters for a fixed receiver. Here, we refer to $\mathbf{R}_{\text{Rx}}$ and $\mathbf{R}_{\text{Tx}}$ collectively as *one-sided correlation matrices*.

Because both $\mathbf{R}_{\text{Rx}}$ and $\mathbf{R}_{\text{Tx}}$ are Hermitian, we can compute their EVD,

$$
\begin{aligned}
\mathbf{R}_{\text{Rx}} &= \sum_{m=1}^{M_{\text{Rx}}} \lambda_{\text{Rx},m} \mathbf{u}_{\text{Rx},m} \mathbf{u}_{\text{Rx},m}^{H} \\
&= \mathbf{U}_{\text{Rx}} \mathbf{\Lambda}_{\text{Rx}} \mathbf{U}_{\text{Rx}}^{H} \\
\mathbf{R}_{\text{Tx}} &= \sum_{n=1}^{M_{\text{Tx}}} \lambda_{\text{Tx},n} \mathbf{u}_{\text{Tx},n} \mathbf{u}_{\text{Tx},n}^{H} \\
&= \mathbf{U}_{\text{Tx}} \mathbf{\Lambda}_{\text{Tx}} \mathbf{U}_{\text{Tx}}^{H},
\end{aligned}
\tag{3.14}
$$

where $\mathbf{u}_{\text{Rx},m}$ and $\mathbf{u}_{\text{Tx},n}$ are the *receive* and *transmit eigenvectors*, and $\mathbf{U}_{\text{Rx}}$ and $\mathbf{U}_{\text{Tx}}$ are the *receive* and *transmit eigenbases*, respectively. We collectively refer to $\mathbf{u}_{\text{Rx},m}$ and $\mathbf{u}_{\text{Tx},n}$ as the *one-sided eigenvectors* and to $\mathbf{U}_{\text{Rx}}$ and $\mathbf{U}_{\text{Tx}}$ as the *one-sided eigenbases*.

3.4 THE KRONECKER MODEL

3.4.1 The Narrowband Kronecker Model

The narrowband Kronecker model (26, 31) is based on the assumption that scatterers around the transmitter are uncorrelated with respect to those around the receiver. In some cases, if the transmitter and receiver are separated by a large distance, both the transmitter and receiver will only be affected by scatterers in their immediate vicinity. Thus, we can visualize two rings formed by the scatterers that most affect the transmitter and receiver. Because of the distance between the rings, the scatterers seen by the transmitter will not correlate with those seen by the receiver. This is illustrated in Fig. 3.1. In this case, the Kronecker model assumes that the transmit-correlation and receive-correlation matrices describe the spatial structure around the transmitter and receiver only. It also assumes that the one-sided correlation matrices are not coupled in any way.

Thus, the only parameters of the Kronecker model are the one-sided correlation matrices. Often, in the literature, the one-sided correlation matrices include

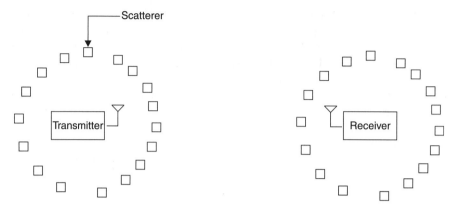

Figure 3.1 Illustration of the two-ring model, where scatterers in the channel can be separated into two rings of scatterers. Each ring describes the scattering around the transmitter and receiver, with no coupling between scatterers at either link end.

normalization constants (71),

$$\mathbf{R}_{\text{Rx,Kron}} = \frac{1}{\beta}\text{E}\{\mathbf{H}\mathbf{H}^H\}$$

$$\mathbf{R}_{\text{Tx,Kron}} = \frac{1}{\alpha}\text{E}\{\mathbf{H}^H\mathbf{H}\}^T, \tag{3.15}$$

where $\mathbf{H}$ has been normalized such that the average receive SNR, averaged across all receive ports, is unity. In this case, the constants α and β satisfy

$$\alpha\beta = \text{Tr}(\mathbf{R}_H) = \text{E}\{\|\mathbf{H}\|_F^2\}, \tag{3.16}$$

where $\text{Tr}(\cdot)$ is the matrix trace operator, defined to be the sum of the diagonal elements. The Kronecker model derives its name from the assumption that the correlation matrix $\mathbf{R}_H$ can be approximated as the Kronecker product of the one-sided correlation matrices; that is,

$$\mathbf{R}_{\text{Kron}} = \mathbf{R}_{\text{Tx,Kron}} \otimes \mathbf{R}_{\text{Rx,Kron}}, \tag{3.17}$$

where the implicit assumption is that $\mathbf{R}_{\text{Kron}} \approx \mathbf{R}_H$. This assumption reduces the number of parameters needed to describe the correlation in the channel from $(M_{\text{Rx}}M_{\text{Rx}})^2$ to $(M_{\text{Rx}}^2 + M_{\text{Tx}}^2)$. An ensemble of H-matrices with the same spatial structure as $\mathbf{H}$ can be synthesized by spatially filtering a statistically white matrix as

$$\mathbf{H}_{\text{Kron}} = \mathbf{R}_{\text{Rx,Kron}}^{1/2}\mathbf{G}(\mathbf{R}_{\text{Tx,Kron}}^{1/2})^T, \tag{3.18}$$

where $\mathbf{R}_{\text{Rx,Kron}}^{1/2}$ and $\mathbf{R}_{\text{Tx,Kron}}^{1/2}$ are the matrix square-roots of $\mathbf{R}_{\text{Rx,Kron}}$ and $\mathbf{R}_{\text{Tx,Kron}}$, respectively, and $\mathbf{G} \in \mathbb{C}^{M_{\text{Rx}} \times M_{\text{Tx}}}$ is populated with complex-Gaussian entries.

3.4.2 The Wideband Kronecker Model

As described in Chapter 2, the wideband H-matrix $\mathbf{H}[d]$ characterizes the time-invariant wideband MIMO channel. We can view each $\mathbf{H}[d]$ for a given d as a narrowband MIMO channel and thus view the wideband MIMO channel as a collection of D narrowband MIMO channels. If we assume that each narrowband MIMO channel is independent of all the others, then we can extend the Kronecker model to the wideband case by applying it to each $\mathbf{H}[d]$ for $d \in \{1, \ldots, D\}$. This approach was first proposed in (36, 72). The resulting synthesis equation is

$$\mathbf{H}_{\mathrm{Kron}}[d] = \mathbf{R}_{\mathrm{Rx}}^{1/2}[d]\mathbf{G}[d](\mathbf{R}_{\mathrm{Tx}}^{1/2}[d])^{T}, \qquad (3.19)$$

where $\mathbf{R}_{\mathrm{Rx}}[d]$ and $\mathbf{R}_{\mathrm{Tx}}[d]$ are computed from $\mathbf{H}[d]$ as

$$\begin{aligned}
\mathbf{R}_{\mathrm{Rx}}[d] &= E\{\mathbf{H}[d]\mathbf{H}^{H}[d]\} \\
\mathbf{R}_{\mathrm{Tx}}[d] &= E\{\mathbf{H}^{T}[d]\mathbf{H}^{*}[d]\}
\end{aligned} \qquad (3.20)$$

for $d \in \{1, \ldots, D\}$. To avoid confusion, in the following, when we use the term *Kronecker model*, we imply the wideband Kronecker model given by (3.19).

3.4.3 Notes on the Narrowband and Wideband Kronecker Models

Detailed analysis of the narrowband and wideband Kronecker models using real-life data have shown that it performs poorly, especially for systems in which the number of antennas is large (>3) (50, 51). In his dissertation (73), Yu quotes model errors as high as 49% and 60% for 3×2 and 2×3 arrays, respectively. In Ref. 50, Özcelik et al. provide an analysis of the MIMO APSs from real-life channel measurements. It was shown therein that, in typical non–line-of-sight (NLoS) channels, the DoD spectra are *not* independent of the DoA spectra. This implies that the scatterers at either end of the link are, in fact, correlated. The Kronecker model forces independence of the two spectra, which produces artifact paths at the vertical and horizontal intersections of real paths that are not present in the actual channel. The artifact paths increase the apparent diversity but decrease the apparent channel capacity. The strength of any given artifact path depends on the number and strength of the real paths intersecting at that point. Because the overall power in the channel is kept constant through normalization, this has the effect of taking power away from real paths in the channel if they do not lie at the intersection of other paths. The end result is that the Kronecker model *consistently underestimates capacity*. Despite the evidence that suggests the Kronecker model to be inaccurate, it is by far the most popular MIMO channel model, mainly due to its simplicity and analytical tractability.

The effects of the separability assumption are best illustrated by way of an exemplar MIMO APS. Consider the example given in Section 2.6.2, where we synthetically generated a correlation matrix $\mathbf{R}_{\mathrm{H,Kron}}$ having four equally weighted and correlated scatterers located at $(\varphi_{\mathrm{Rx},s}, \varphi_{\mathrm{Tx},s}) = (60°, 40°), (0°, -50°), (-50°, 60°), (0°, 0°)$. To test the Kronecker model's ability to approximate the spatial structure of a given

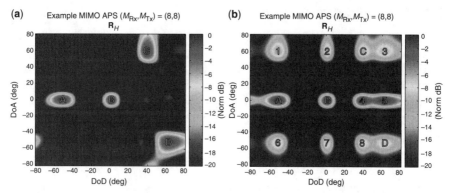

Figure 3.2 The effects of the separability assumption on the MIMO APS. (a) MIMO APS of the true channel, showing four resolvable scatterers labeled A–D. (b) MIMO APS using the Kronecker model, showing artifact paths labeled 1–8.

channel, (3.7) is used to generate an ensemble of H-matrices. Equation (3.15) is used to compute the one-sided correlation matrices $\mathbf{R}_{\text{Rx,Kron}}$ and $\mathbf{R}_{\text{Tx,Kron}}$. Finally, (3.17) is used to compute the full correlation matrix $\mathbf{R}_{\text{Kron}}$, which is then used to compute the Kronecker APS. Figure 3.2a shows the true MIMO APS, computed using $\mathbf{R}_H$, and Fig. 3.2b shows the Kronecker APS.

In Fig. 3.2b, the Kronecker APS includes the four original paths, labeled A–D, and eight artifact paths, labeled 1–8. Each artifact path occurs at the intersection of real paths. Artifact paths 4 and 5 are particularly strong, as they are both products of *three* real paths. Note that the magnitudes of real paths C and D are greatly diminished as they do not lie at the intersection points of any other real paths. In general, increasing the number of resolved scatterers will increase the number of artifact paths and tends to diminish energy in paths that depart/arrive farther from the array boresight. This means that the Kronecker model performance degrades as we increase spatial resolution. We can increase spatial resolution, for example, by increasing the number of antennas and/or antenna directivity.

It is reasonable to ask under what circumstances the Kronecker model might perform accurately. Given the above discussion, we hypothesize that the Kronecker model will perform better in cases where the scattering is not correlated across the link and there is a system with relatively few antennas (namely 2 × 2). For example, consider the case where the transmitter is separated from the receiver by a considerable distance. Let the transmitter reside above the local clutter height, so that there is no significant scattering around the transmitter. Let the receiver reside in an area with good local scattering. This situation approximately describes that of a base station–mobile scenario. In this case, it is hypothesized that the scatterers would not be correlated between link-ends and that the Kronecker model might perform reasonably well.

The above analysis begs the question, "How often does this happen in real-life channels?" We discuss this issue further in Chapter 6, using MIMO APSs computed from real-life data, and compare the results with those of the Kronecker and structured models.

3.5 THE WEICHSELBERGER MODEL

In his thesis (51), Weichselberger presents two novel correlative channel models based on approximations of $\mathbf{R}_H$. The first model is based on the concept of *vector modes*, which he defines in his thesis. We will refer to the resulting model as the *vector mode model* and provide a brief overview in the following. The second model is based on another novel concept, which he calls *structured modes*. This model is the version most often presented in the literature. We therefore refer to the last model as the *Weichselberger model*.

The Weichselberger model is based on the assumption that scatterers at each link end are coupled. According to the model, the EVD is used to transform the one-sided correlation matrices to the eigendomain. The *coupling coefficients* are computed as the average power coupled between the eigenbases of the one-sided correlation matrices. Thus, the parameters of the Weichselberger model are the eigenbases of the one-sided correlation matrices and the average power coupled between eigenvectors. The following sections contain a brief overview of the vector mode and Weichselberger models.

3.5.1 The Vector Mode Model

Consider the EVD of the narrowband correlation matrix, repeated here for convenience of presentation:

$$\mathbf{R}_H = \sum_{k=1}^{M_{Rx}M_{Tx}} \lambda_k \mathbf{u}_k \mathbf{u}_k^H. \tag{3.21}$$

The elements of the eigenvectors $\mathbf{u}_k$ are used to form the eigenbasis $\mathbf{U}_k$. From the above discussion, rank $(\mathbf{U}_k)$ is equal to the number of correlated scatterers for the kth eigenvalue. In general, in physical channels, the likelihood that two separately resolvable scatterers are correlated is extremely low. Scatterers that can be resolved in space are usually separated by a considerable distance and thus are usually uncorrelated. Because of this fact, Weichselberger bases the vector mode model on the assumption that each $\mathbf{U}_k$ can be approximated as a rank-one matrix. This, in turn, means that each $M_{Rx} M_{Tx} \times 1$ eigenvector $\mathbf{u}_k$ can be approximated as the Kronecker product of two vectors, say an $M_{Tx} \times 1$ vector $\boldsymbol{\eta}_k$ and an $M_{Rx} \times 1$ vector $\mathbf{v}_k$. By doing this, the number of parameters needed to describe the EVD of the correlation matrix is reduced, but the approximation introduces some finite error. The vectors $\mathbf{v}_k$ and $\boldsymbol{\eta}_k$ are called the *left* and *right vector modes* of the channel. The vector modes are defined such that

1. The decomposition of the channel using vector modes is unique.
2. We get non-negative power terms (vector mode values, similar to eigenvalues).
3. The modes must be Hermitian symmetric.
4. The modes must be mutually orthogonal.

Given these constraints, the correlation matrix can be approximated as

$$\mathbf{R}_H \approx \mathbf{R}_{\text{vec}} = \sum_{k=1}^{M_{\text{Rx}}M_{\text{Tx}}} \mu_k (\boldsymbol{\eta}_k \otimes \boldsymbol{v}_k)(\boldsymbol{\eta}_k \otimes \boldsymbol{v}_k)^H, \tag{3.22}$$

where μ_k is the kth *vector mode value*.

Comparing (3.21) and (3.22), we see that the vector modes are constructed such that they resemble the eigenvectors of the channel. The vector modes and vector mode values are computed from the measured correlation matrix using optimization techniques. The relevant equations and resulting algorithm are complex and will not be repeated here. Because vector modes approximate the eigenbases of the channel as rank-one matrices, they cannot model the correlation between scatterers in each eigenbasis. This introduces modeling error.

Using the above approximation for $\mathbf{R}_H$, we can recast (3.8) in terms of vector modes as follows:

$$\mathbf{H}_{\text{vec}} = \sum_{k=1}^{M_{\text{Rx}}M_{\text{Tx}}} g_k \sqrt{\mu_k} \, \boldsymbol{v}_k \boldsymbol{\eta}_k^T, \tag{3.23}$$

where g_k are complex-Gaussian random variables with zero mean and unit variance. The *vector mode synthesis equation* (3.23) reduces the number of parameters needed to synthesize an exemplar H-matrix versus (3.8), from $(M_{\text{Rx}}M_{\text{Tx}})^2 + M_{\text{Rx}}M_{\text{Tx}}$ to $(M_{\text{Rx}}M_{\text{Tx}})(M_{\text{Rx}} + M_{\text{Tx}}) + M_{\text{Rx}}M_{\text{Tx}}$.

3.5.2 H-matrix from Structured Vector Modes

Note that vector modes do not have any physical structure; that is, the elements in the vector modes cannot be related to any single antenna or antenna pair. This precludes its use in formulating signal strategies at either link end. As a result, the vector mode model does not give any insight into the behavior of the channel at either link end. Also, the vector modes are, in general, difficult to compute: The left and right vector modes $\boldsymbol{v}_k$ and $\boldsymbol{\eta}_k^T$ and the vector mode values μ_k are computed using an iterative optimization algorithm. To bring some structure to the channel model, Weichselberger places restrictions on the space from which we can choose the vector modes (51).

To this end, from Section 3.3, the one-sided correlation matrices are computed as

$$\mathbf{R}_{\text{Rx}} = \mathrm{E}\{\mathbf{H}\mathbf{H}^H\}$$
$$\mathbf{R}_{\text{Tx}} = \mathrm{E}\{\mathbf{H}^T\mathbf{H}^*\}. \tag{3.24}$$

Also recall that the one-sided eigenvectors from the EVD of the one-sided correlation matrices can be computed as

$$\mathbf{R}_{\text{Rx}} = \sum_{m=1}^{M_{\text{Rx}}} \lambda_{\text{Rx},m} \mathbf{u}_{\text{Rx},m} \mathbf{u}_{\text{Rx},m}^H$$
$$\mathbf{R}_{\text{Tx}} = \sum_{n=1}^{M_{\text{Tx}}} \lambda_{\text{Tx},n} \mathbf{u}_{\text{Tx},n} \mathbf{u}_{\text{Tx},n}^H. \tag{3.25}$$

Weichselberger brings structure to the vector mode model by restricting the choice of vector modes to the space spanned by the one-sided eigenvectors; that is,

$$\begin{aligned} \boldsymbol{v}_k &\in \mathfrak{A}_{\text{Rx}} \\ \boldsymbol{\eta}_k &\in \mathfrak{A}_{\text{Tx}}, \end{aligned} \tag{3.26}$$

where $\mathfrak{A}_{\text{Rx}}$ and $\mathfrak{A}_{\text{Tx}}$ are the space spanned by all $\mathbf{u}_{\text{Rx},m}$ and $\mathbf{u}_{\text{Tx},n}$, respectively. The vector mode synthesis equation (3.23) can be reformulated using the one-sided eigenvectors as

$$\mathbf{H}_{\text{Weich}} = \sum_{m=1}^{M_{\text{Rx}}} \sum_{n=1}^{M_{\text{Tx}}} g_{mn} \sqrt{\omega_{mn}} \mathbf{u}_{\text{Rx},m} \mathbf{u}_{\text{Tx},n}^H, \tag{3.27}$$

where ω_{mn} are the *structured eigenvalues*. The structured eigenvalues are defined as the average power coupled between the mth receive and the nth transmit eigenvector; that is,

$$\begin{aligned} \omega_{mn} &= \text{E}\{\|\mathbf{u}_{\text{Rx},m}^H \mathbf{H} \mathbf{u}_{\text{Tx},n}^*\|^2\} \\ &= \text{E}\{[(\mathbf{u}_{\text{Tx},n} \otimes \mathbf{u}_{\text{Rx},m})^H \text{vec}(\mathbf{H})]^H [(\mathbf{u}_{\text{Tx},n} \otimes \mathbf{u}_{\text{Rx},m})^H \text{vec}(\mathbf{H})]\} \\ &= (\mathbf{u}_{\text{Tx},n} \otimes \mathbf{u}_{\text{Rx},m})^H \mathbf{R}_H (\mathbf{u}_{\text{Tx},n} \otimes \mathbf{u}_{\text{Rx},m}). \end{aligned} \tag{3.28}$$

For this reason, ω_{mn} is also referred to as the *coupling coefficient*. In the above derivation, we make use of the identity

$$\text{vec}(\mathbf{AXB}) = (\mathbf{B}^T \otimes \mathbf{A})\text{vec}(\mathbf{X}), \tag{3.29}$$

and the fact that $\mathbf{u}_{\text{Rx},m}^H \mathbf{H} \mathbf{u}_{\text{Tx},n}^*$ is a scalar quantity. We can arrange the ω_{mn} in matrix form as

$$\boldsymbol{\Omega} = \begin{pmatrix} \omega_{11} & \cdots & \omega_{1M_{\text{Tx}}} \\ \vdots & \ddots & \vdots \\ \omega_{M_{\text{Tx}}1} & \cdots & \omega_{M_{\text{Rx}}M_{\text{Tx}}} \end{pmatrix}, \tag{3.30}$$

where $\boldsymbol{\Omega}$ is the *coupling matrix*. Using the above relations, $\mathbf{H}_{\text{Weich}}$ is recast more compactly as

$$\mathbf{H}_{\text{Weich}} = \mathbf{U}_{\text{Rx}}(\widetilde{\boldsymbol{\Omega}} \odot \mathbf{G})\mathbf{U}_{\text{Tx}}^T, \tag{3.31}$$

where

$$\begin{aligned} \mathbf{U}_{\text{Rx}} &= \begin{bmatrix} \mathbf{u}_{\text{Rx},1} & \cdots & \mathbf{u}_{\text{Rx},M_{\text{Rx}}} \end{bmatrix} \\ \mathbf{U}_{\text{Tx}} &= \begin{bmatrix} \mathbf{u}_{\text{Tx},1} & \cdots & \mathbf{u}_{\text{Tx},M_{\text{Tx}}} \end{bmatrix}, \end{aligned} \tag{3.32}$$

are the one-sided eigenvectors, $\widetilde{\mathbf{\Omega}}$ is the element-wise square root of $\mathbf{\Omega}$ such that

$$
\widetilde{\mathbf{\Omega}} = \begin{pmatrix} \sqrt{\omega_{11}} & \cdots & \sqrt{\omega_{1M_{\text{Tx}}}} \\ \vdots & \ddots & \vdots \\ \sqrt{\omega_{M_{\text{Tx}}1}} & \cdots & \sqrt{\omega_{M_{\text{Rx}}M_{\text{Tx}}}} \end{pmatrix}, \tag{3.33}
$$

$\mathbf{G} \in \mathbb{C}^{M_{\text{Rx}} \times M_{\text{Tx}}}$ has complex-Gaussian elements, and $\odot$ is the Hadamard product. We refer to (3.31) as the *Weichselberger model synthesis equation*.

The Weichselberger model has some nice properties versus the vector mode model. First, its parameters are relatively easy to compute compared with those of the vector mode model. Second, we can compute the parameters directly from the EVD of the one-sided correlation matrices. Moreover, the structure of $\mathbf{\Omega}$ tells us a lot about the spatial structure of the channel. This facilitates the development of signaling strategies at both link ends. We discuss coupling coefficients, and their relation to the orientation of scatterers in the physical channel, in more detail in the following section. The synthesis equation (3.31) also uses fewer parameters than the vector mode model and far fewer parameters than the full correlation matrix synthesis equation (3.7). The downside is that it is less accurate than the above-mentioned models.

The main disadvantage of the Weichselberger versus Kronecker model is the increased number of parameters needed to synthesize an exemplar H-matrix. The Weichselberger model requires $M_{\text{Rx}}M_{\text{Tx}} + M_{\text{Rx}}(M_{\text{Rx}} - 1) + M_{\text{Tx}}(M_{\text{Tx}} - 1)$ parameters, whereas the Kronecker model requires $M_{\text{Rx}}^2 + M_{\text{Tx}}^2$. Despite this disadvantage, the Weichselberger model has been shown to closely agree with measured data, especially with respect to predicting the capacity of the channel. Using real-life data, the Weichselberger model has been shown to consistently outperform the Kronecker model (51, 74).

It is interesting to note that the Weichselberger model can be viewed as a generalized version of the Kronecker model. The Weichselberger model reduces to the Kronecker model if, and only if, the coupling matrix reduces to the outer product of the transmit- and receive-eigenvalues; that is, if

$$
\mathbf{\Omega} = \frac{1}{P_{\text{Rx}}} \begin{pmatrix} \lambda_{\text{Rx}1} \\ \lambda_{\text{Rx}2} \\ \vdots \\ \lambda_{\text{Rx}N} \end{pmatrix} \begin{pmatrix} \lambda_{\text{Tx}1} \\ \lambda_{\text{Tx}2} \\ \vdots \\ \lambda_{\text{Tx}N} \end{pmatrix}^{T}. \tag{3.34}
$$

The above condition implies that the coupling matrix is of rank one.

3.6 THE STRUCTURED MODEL

This section introduces the third correlative wideband MIMO channel model, called *the structured model*. The structured model derives its name from the fact that it is based on the concept of structured vector modes, extended to include the wideband case. The structured model uses the correlation between paths to approximate the *spatial structure* of a *wideband MIMO* channel. The model considers correlation in three dimensions: across the transmitter, receiver, and delay spaces; thus it preserves more of the structure of the wideband MIMO channel than does the Kronecker model. In Chapter 6, using real-life data, we show that the structured model consistently outperforms the Kronecker model.

First, the ideas presented in Section 3.2 are extended to develop a new synthesis equation that uses the wideband correlation matrix $\mathbf{R}_{WB,H}$ to generate an ensemble of H-tensors. The concept of one-sided correlation is extended to the wideband case using tensor calculus. The concept of *wideband coupling coefficients* is defined using the EVD of the one-sided correlation matrices. This leads naturally to the development of the structured model synthesis equation.

3.6.1 H-tensor Synthesis from the Wideband Correlation Tensor

In the following, we wish to derive a synthesis equation in which we use the full wideband correlation matrix $\mathbf{R}_{WB,H}$ to generate an ensemble of H-tensors. This is done by extending (3.7) to include the wideband case. The result is that

$$\mathcal{H}_{\text{synth}} = \text{unvec}(\mathbf{R}_{WB,H}^{1/2}\mathbf{g}_H), \tag{3.35}$$

where $\mathbf{g}_H \in \mathbb{C}^{M_{Rx}M_{Tx}D \times 1}$ is a complex-Gaussian vector, and unvec($\cdot$) operates on all three dimensions. The unvec($\cdot$) operator, applied to tensors, was first defined in Chapter 2. The subscript synth is used to emphasize the fact that $\mathcal{H}_{\text{synth}}$ is synthesized from an ideal distribution. Note that the number of parameters implied by the above model is $(M_{Rx}M_{Tx}D)^2$.

To recast the synthesis equation using the EVD of $\mathbf{R}_{WB,H}$, the $M_{Rx}M_{Tx}D \times 1$ synthetic channel vector $\mathbf{h}_{WB,\text{synth}}$ is defined such that $\mathbf{h}_{WB,\text{synth}} = \text{vec}(\mathcal{H}_{\text{synth}})$. Using this result, (3.35) can be rewritten as

$$\mathbf{h}_{WB,\text{synth}} = \sum_{k=1}^{M_{Rx}M_{Tx}D} g_k\lambda_{WB,k}^{1/2}\mathbf{u}_{WB,k}. \tag{3.36}$$

In the following, we show how the above-proposed model maintains the spatial structure of the given channel. Taking the expected value of the complex outer product of $\mathbf{h}_{WB,\text{synth}}$ with itself gives

$$\mathbf{R}_{WB,H} = \text{E}\{\mathbf{h}_{WB,\text{synth}}\mathbf{h}_{WB,\text{synth}}^H\}$$

$$= \text{E}\left\{\sum_k\sum_l g_k\lambda_{WB,k}^{1/2}\mathbf{u}_{WB,k}(g_l\lambda_{WB,l}^{1/2}\mathbf{u}_{WB,l})^H\right\}$$

$$= \sum_k\sum_l \lambda_{WB,k}^{1/2}\lambda_{WB,l}^{1/2}\mathbf{u}_{WB,k}\mathbf{u}_{WB,l}^H\text{E}\{g_kg_l^*\}$$

$$= \sum_k \lambda_{WB,k}\mathbf{u}_{WB,k}\mathbf{u}_{WB,k}^H, \tag{3.37}$$

where the last line results from the fact that $E\{g_k g_l^*\} = 0$ for $k \neq l$. Thus, we have shown that (3.35) preserves the spatial structure of the channel and that we can synthesize $\mathcal{H}_{\text{synth}}$ from the EVD of $\mathbf{R}_{\text{WB},H}$.

3.6.2 One-Sided Correlation for Wideband MIMO Channels

The concept of one-sided correlation can be extended to the wideband case. This imposes more structure on the channel than that implied by (3.35). First, note that $\mathbf{U}_{\text{Rx}}$ and $\mathbf{U}_{\text{Tx}}$, given in (3.14), are the bases for the row (receiver) and column (transmitter) space of $\mathbf{H}$, respectively. To extend this to the wideband case, we need a way to generate an orthonormal basis for each dimension of the H-tensor. From Appendix A, the *higher-order singular value decomposition* (HOSVD) can be used to decompose the H-tensor into an all-orthogonal core tensor and three orthogonal bases, viz. (75),

$$\mathcal{H} = \mathcal{S}_H \times_1 \mathbf{U}_1 \times_2 \mathbf{U}_2 \times_3 \mathbf{U}_3, \tag{3.38}$$

where $\mathcal{S}_H \in \mathbb{C}^{I_1 \times I_2 \times I_3}$ is the core tensor, $\times_n$ denotes the n-mode product, and $\mathbf{U}_n$ is an orthogonal bases for each dimension of $\mathcal{H}$. It is important to note that the HOSVD of H does not describe the average behavior of the channel, nor does it describe its spatial structure. This, in turn, implies that it cannot be used directly to synthesize new channels with the same spatial structure. To derive the structured model, we need to work with the channel correlation.

Let $\mathbf{H}_{(n)}$ be the nth unfolding of $\mathcal{H}$. We compute the nth orthogonal basis $\mathbf{U}_n$ as the left singular vector of the nth unfolding of $\mathcal{H}$; that is,

$$\mathbf{H}_{(n)} = \mathbf{U}_n \mathbf{S}_n \mathbf{V}_n^T. \tag{3.39}$$

The $\mathbf{H}_{(n)}$ can be viewed as the mapping of the elements h_{mnd}, where m, n, and d refer to the indices of the first, second, and third dimensions of $\mathcal{H}$, respectively, to a matrix such that the nth dimension varies across the row space, and the other dimensions vary across the column space. For example, the first matrix unfolding, $\mathbf{H}_{(1)}$, has the following structure,

$$\mathbf{H}_{(1)} = \begin{pmatrix} h_{111} & \cdots & h_{11D} & h_{121} & \cdots & h_{12D} & \cdots & h_{1M_{\text{Tx}}1} & \cdots & h_{1M_{\text{Tx}}D} \\ \vdots & & \vdots & \vdots & & \vdots & & \vdots & & \vdots \\ h_{M_{\text{Rx}}11} & \cdots & h_{M_{\text{Rx}}1D} & h_{M_{\text{Rx}}21} & \cdots & h_{M_{\text{Rx}}2D} & \cdots & h_{M_{\text{Rx}}M_{\text{Tx}}1} & \cdots & h_{M_{\text{Rx}}M_{\text{Tx}}D} \end{pmatrix}. \tag{3.40}$$

In this way, any given column of $\mathbf{H}_{(1)}$ is similar to $\mathbf{h}_{\text{col},n}$, first defined in (3.11), in that it describes the SIMO channel for all receivers given a fixed transmitter and delay.

Similarly, the second matrix unfolding of $\mathcal{H}$ has the structure

$$
\mathbf{H}_{(2)} = \begin{pmatrix} h_{111} & \cdots & h_{M_{\mathrm{Rx}}11} & h_{112} & \cdots & h_{M_{\mathrm{Rx}}12} & \cdots & h_{11D} & \cdots & h_{M_{\mathrm{Rx}}1D} \\ \vdots & & \vdots & \vdots & & \vdots & & \vdots & & \vdots \\ h_{1M_{\mathrm{Tx}}1} & \cdots & h_{M_{\mathrm{Rx}}M_{\mathrm{Tx}}1} & h_{1M_{\mathrm{Tx}}2} & \cdots & h_{M_{\mathrm{Rx}}M_{\mathrm{Tx}}2} & \cdots & h_{1M_{\mathrm{Tx}}D} & \cdots & h_{M_{\mathrm{Rx}}M_{\mathrm{Tx}}D} \end{pmatrix}.
$$
(3.41)

Any column of $\mathbf{H}_{(2)}$ is similar to $\mathbf{h}_{\mathrm{row},m}$, defined in (3.10), in that it can be viewed as the MISO channel formed by considering all transmitters and keeping the receiver and delay fixed. It follows that the columns of the third matrix unfolding $\mathbf{H}_{(3)}$ are the vector channels formed by taking all h_{mnd} for a fixed delay d and allowing m and n to vary across all receivers and transmitters, respectively.

Using the same arguments as for the narrowband case in Section 3.3, the definition of one-sided correlation can be extended to the H-tensor. The *receive-correlation matrix* $\mathbf{R}_{\mathrm{Rx}} \in \mathbb{C}^{M_{\mathrm{Rx}} \times M_{\mathrm{Rx}}}$ is defined as

$$
\mathbf{R}_{\mathrm{Rx}} = \mathrm{E}\{\mathbf{H}_{(1)}\mathbf{H}_{(1)}^H\},
$$
(3.42)

and the *transmit-correlation matrix* $\mathbf{R}_{\mathrm{Tx}} \in \mathbb{C}^{M_{\mathrm{Tx}} \times M_{\mathrm{Tx}}}$ is correspondingly defined as

$$
\mathbf{R}_{\mathrm{Tx}} = \mathrm{E}\{\mathbf{H}_{(2)}\mathbf{H}_{(2)}^H\}.
$$
(3.43)

Finally, the *delay-correlation matrix* $\mathbf{R}_{\mathrm{Del}} \in \mathbb{C}^{D \times D}$ is defined to be

$$
\mathbf{R}_{\mathrm{Del}} = \mathrm{E}\{\mathbf{H}_{(3)}\mathbf{H}_{(3)}^H\}.
$$
(3.44)

Note that, for the narrowband case, $\mathbf{H}$ can be considered to be a second-order tensor, and

$$
\begin{aligned}
\mathrm{E}\{\mathbf{H}_{(1)}\mathbf{H}_{(1)}^H\} &= \mathrm{E}\{\mathbf{H}\mathbf{H}^H\} \\
\mathrm{E}\{\mathbf{H}_{(2)}\mathbf{H}_{(2)}^H\} &= \mathrm{E}\{\mathbf{H}^H\mathbf{H}^*\},
\end{aligned}
$$
(3.45)

which reflects (3.24).

The EVD can be applied to the one-sided correlation matrices to obtain a basis for the receiver, transmitter, and delay space, as follows, respectively:

$$
\mathbf{R}_{\mathrm{Rx}} = \sum_{i=1}^{M_{\mathrm{Rx}}} \lambda_{\mathrm{Rx},i}\mathbf{u}_{\mathrm{Rx},i}\mathbf{u}_{\mathrm{Rx},i}^H = \mathbf{U}_{\mathrm{Rx}}\mathbf{\Lambda}_{\mathrm{Rx}}\mathbf{U}_{\mathrm{Rx}}^T
$$

$$\mathbf{R}_{\mathrm{Tx}} = \sum_{j=1}^{M_{\mathrm{Tx}}} \lambda_{\mathrm{Tx},j} \mathbf{u}_{\mathrm{Tx},j} \mathbf{u}_{\mathrm{Tx},j}^{H} = \mathbf{U}_{\mathrm{Tx}} \mathbf{\Lambda}_{\mathrm{Rx}} \mathbf{U}_{\mathrm{Tx}}^{H}$$

$$\mathbf{R}_{\mathrm{Del}} = \sum_{k=1}^{D} \lambda_{\mathrm{Del},k} \mathbf{u}_{\mathrm{Del},k} \mathbf{u}_{\mathrm{Del},k}^{H} = \mathbf{U}_{\mathrm{Del}} \mathbf{\Lambda}_{\mathrm{Del}} \mathbf{U}_{\mathrm{Del}}^{H}.$$

(3.46)

We collectively refer to the matrices $\mathbf{U}_{\mathrm{Rx}}$, $\mathbf{U}_{\mathrm{Tx}}$, and $\mathbf{U}_{\mathrm{Del}}$ as the *one-sided eigenbases*.

3.6.3 Approximating the Wideband Correlation Matrix

To reduce the number of parameters needed to synthesize an H-tensor, the eigenbases of the wideband correlation matrix $\mathbf{R}_{\mathrm{WB},H}$ may be approximated using the one-sided eigenbases $\mathbf{U}_{\mathrm{Rx}}$, $\mathbf{U}_{\mathrm{Tx}}$, and $\mathbf{U}_{\mathrm{Del}}$ first defined in Section 3.6.2. Consider the EVD of $\mathbf{R}_{\mathrm{WB},H}$ repeated here for clarity,

$$\mathbf{R}_{\mathrm{WB},H} = \sum_{k=1}^{M_{\mathrm{Rx}}M_{\mathrm{Tx}}D} \lambda_{\mathrm{WB},k} \mathbf{u}_{\mathrm{WB},k} \mathbf{u}_{\mathrm{WB},k}^{H}.$$

(3.47)

Note that $\mathbf{u}_{\mathrm{WB},k}$, for each k, consists of $M_{\mathrm{Rx}}M_{\mathrm{Tx}}D$ elements.

Extending the ideas from the Weichselberger model, $\mathbf{R}_{\mathrm{WB},H}$ can be approximated by the Kronecker product of three vectors of dimension $M_{\mathrm{Rx}} \times 1$, $M_{\mathrm{Tx}} \times 1$, and $D \times 1$. Furthermore, let these vectors be chosen from the set of one-sided eigenvectors $\mathbf{u}_{\mathrm{Rx},i}$, $\mathbf{u}_{\mathrm{Tx},j}$, and $\mathbf{u}_{\mathrm{Del},k}$. In this way, we can approximate $\mathbf{R}_{\mathrm{WB},H}$ as the triple summation:

$$\mathbf{R}_{\mathrm{WB},H} \approx \mathbf{R}_{\mathrm{WB,struct}} = \sum_{i=1}^{M_{\mathrm{Rx}}} \sum_{j=1}^{M_{\mathrm{Tx}}} \sum_{k=1}^{D} \omega_{ijk} (\mathbf{u}_{\mathrm{Del},k} \otimes \mathbf{u}_{\mathrm{Tx},j} \otimes \mathbf{u}_{\mathrm{Rx},i})$$

$$\times (\mathbf{u}_{\mathrm{Del},k} \otimes \mathbf{u}_{\mathrm{Tx},j} \otimes \mathbf{u}_{\mathrm{Rx},i})^{H},$$

(3.48)

where $\mathbf{u}_{\mathrm{Rx},i}$, $\mathbf{u}_{\mathrm{Tx},j}$, and $\mathbf{u}_{\mathrm{Del},k}$ are the one-sided eigenvectors, and ω_{ijk} are the *wideband coupling coefficients*, computed as the average energy coupled between one-sided eigenvectors. We refer to $\mathbf{R}_{\mathrm{WB,struct}}$ as the *structured wideband correlation matrix*.

By approximating the eigenvector $\mathbf{u}_{\mathrm{WB},k}$ as the Kronecker product of three vectors, the number of parameters needed for each vector has been reduced from $M_{\mathrm{Rx}}M_{\mathrm{Tx}}D$ to $M_{\mathrm{Rx}} + M_{\mathrm{Tx}} + D$, which is practically significant. The following section outlines the method by which we compute the wideband coupling coefficients from the one-sided eigenvectors. We begin with the narrowband case and then extend the definition to include the wideband case.

3.6.3.1 *The Narrowband Case* Before defining the wideband coupling coefficient, it is instructive to revisit Weichselberger's definition of a coupling coefficient, given by (3.28), and restated here for clarity. Given an $M_{\mathrm{Tx}} \times 1$ transmit

eigenvector $\mathbf{u}_{\text{Tx},j}$, and an $M_{\text{Rx}} \times 1$ receive eigenvector $\mathbf{u}_{\text{Rx},i}$, the power coupled between them through the channel was defined as

$$\omega_{ij} \triangleq \mathrm{E}\{\|\mathbf{u}_{\text{Rx},i}^{H}\mathbf{H}\mathbf{u}_{\text{Tx},j}^{*}\|^{2}\}. \tag{3.49}$$

Let the structure of the narrowband transmit, receive eigenvectors be

$$\mathbf{u}_{\text{Tx},j} = \begin{bmatrix} u_{\text{Tx},1j} & u_{\text{Tx},2j} & \cdots & u_{\text{Tx},M_{\text{Tx}}j} \end{bmatrix}^{T}$$
$$\mathbf{u}_{\text{Rx},i} = \begin{bmatrix} u_{\text{Rx},1i} & u_{\text{Rx},2i} & \cdots & u_{\text{Rx},M_{\text{Rx}}i} \end{bmatrix}^{T}. \tag{3.50}$$

The elements of the transmit eigenvector are $u_{\text{Tx},nj}$, where $j \in \{1,\ldots,M_{\text{Tx}}\}$ and $n \in \{1,\ldots,M_{\text{Tx}}\}$. Similarly, the elements of the receive eigenvector are $u_{\text{Rx},mi}$, where $i \in \{1,\ldots,M_{\text{Rx}}\}$ and $m \in \{1,\ldots,M_{\text{Rx}}\}$.

We may gain more insight into the structure of each ω_{ij} by focusing on the inner product term $\mathbf{u}_{\text{Rx},i}^{H}\mathbf{H}\mathbf{u}_{\text{Tx},j}^{*}$. Written explicitly, the inner product term is

$$\mathbf{u}_{\text{Rx},i}^{H}\mathbf{H}\mathbf{u}_{\text{Tx},j}^{*} = \begin{pmatrix} u_{\text{Rx},1i} \\ u_{\text{Rx},2i} \\ \vdots \\ u_{\text{Rx},M_{\text{Rx}}i} \end{pmatrix}^{H} \begin{pmatrix} h_{11} & \cdots & h_{1M_{\text{Tx}}} \\ \vdots & \ddots & \vdots \\ h_{M_{\text{Rx}}1} & \cdots & h_{M_{\text{Rx}}M_{\text{Tx}}} \end{pmatrix} \begin{pmatrix} u_{\text{Tx},1j} \\ u_{\text{Tx},2j} \\ \vdots \\ u_{\text{Tx},M_{\text{Tx}}j} \end{pmatrix}^{*}$$

$$= \sum_{m,n} h_{mn} u_{\text{Rx},mi}^{*} u_{\text{Tx},nj}^{*}, \tag{3.51}$$

with the result being a scalar. Rewritting (3.49) using the above expansion, we write

$$\omega_{ij} = \mathrm{E}\left\{ \left(\sum_{m,n} h_{mn} u_{\text{Rx},mi}^{*} u_{\text{Tx},nj}^{*} \right)^{*} \left(\sum_{p,q} h_{pq} u_{\text{Rx},pi}^{*} u_{\text{Tx},qj}^{*} \right) \right\}. \tag{3.52}$$

In the following, (3.51) and (3.52) are used to extend the definition of a coupling coefficient to the wideband case.

3.6.3.2 The Wideband Case and the Structured Model Synthesis Equation
We next wish to rewrite the inner product term formulated above using three dimensions: transmit, receive, and delay space. To this end, let the inner product of all elements of the one-sided eigenvectors with the corresponding elements from $\mathcal{H}$ be (see Appendix B)

$$\sum_{m=1}^{M_{\text{Rx}}}\sum_{n=1}^{M_{\text{Tx}}}\sum_{d=1}^{D} h_{mnd} u_{\text{Rx},mi}^{*} u_{\text{Tx},nj}^{*} u_{\text{Del},dk}^{*} = (\mathbf{u}_{\text{Del},k} \otimes \mathbf{u}_{\text{Tx},j} \otimes \mathbf{u}_{\text{Rx},i})^{H} \text{vec}(\mathcal{H}). \tag{3.53}$$

The *wideband coupling coefficient* ω_{ijk} is defined as the expected value of this inner product, viz.,

$$\omega_{ijk} = \mathrm{E}\left\{\left(\sum_{m=1}^{M_{\mathrm{Rx}}}\sum_{n=1}^{M_{\mathrm{Tx}}}\sum_{d=1}^{D} h_{mnd} u_{\mathrm{Rx},mi}^* u_{\mathrm{Tx},nj}^* u_{\mathrm{Del},dk}^*\right)^*\left(\sum_{p=1}^{M_{\mathrm{Rx}}}\sum_{q=1}^{M_{\mathrm{Tx}}}\sum_{r=1}^{D} h_{pqr} u_{\mathrm{Rx},pi}^* u_{\mathrm{Tx},qj}^* u_{\mathrm{Del},rk}^*\right)\right\}$$

$$= (\mathbf{u}_{\mathrm{Del},k}\otimes\mathbf{u}_{\mathrm{Tx},j}\otimes\mathbf{u}_{\mathrm{Rx},i})^H \mathbf{R}_{\mathrm{WB},H}(\mathbf{u}_{\mathrm{Del},k}\otimes\mathbf{u}_{\mathrm{Tx},j}\otimes\mathbf{u}_{\mathrm{Rx},i}). \tag{3.54}$$

As in the narrowband case, we can use this result to approximate $\mathbf{R}_{\mathrm{WB},H}$. The EVD of $\mathbf{R}_{\mathrm{WB},H}$ is computed as

$$\mathbf{R}_{\mathrm{WB},H} = \sum_{k=1}^{M_{\mathrm{Rx}}M_{\mathrm{Tx}}D} \lambda_{\mathrm{WB},k}\mathbf{u}_{\mathrm{WB},k}\mathbf{u}_{\mathrm{WB},k}^H. \tag{3.55}$$

Using the above definition for ω_{ijk}, $\mathbf{R}_{\mathrm{WB},H}$ is approximated as

$$\mathbf{R}_{\mathrm{WB},H} \approx \mathbf{R}_{\mathrm{WB,struct}} = \sum_{i=1}^{M_{\mathrm{Rx}}}\sum_{j=1}^{M_{\mathrm{Tx}}}\sum_{k=1}^{D} \omega_{ijk}(\mathbf{u}_{\mathrm{Del},k}\otimes\mathbf{u}_{\mathrm{Tx},j}\otimes\mathbf{u}_{\mathrm{Rx},i})$$

$$\cdot (\mathbf{u}_{\mathrm{Del},k}\otimes\mathbf{u}_{\mathrm{Tx},j}\otimes\mathbf{u}_{\mathrm{Rx},i})^H. \tag{3.56}$$

Finally, it can be shown that the synthesis formula for an H-tensor using the one-sided eigenvectors is

$$\mathcal{H}_{\mathrm{struct}} = \mathcal{G}\times_1\mathbf{U}_{\mathrm{Rx}}\times_2\mathbf{U}_{\mathrm{Tx}}\times_3\mathbf{U}_{\mathrm{Del}}, \tag{3.57}$$

where $\mathcal{G}\in\mathbb{C}^{M_{\mathrm{Rx}}\times M_{\mathrm{Tx}}\times D}$ is the tensor whose elements $w_{mnd} = g_{mnd}\sqrt{\omega_{mnd}}$, and g_{mnd} is a complex-Gaussian random variable. We refer to (3.57) as the *structured model synthesis equation*. The H-tensor $\mathcal{H}_{\mathrm{struct}}$ maintains approximately the same spatial structure as that characterized by $\mathcal{R}$. However, as we see in the next section, the number of parameters needed to generate an exemplar H-tensor is greatly reduced versus that in (3.35).

3.6.4 Number of Parameters Comparison

Section 3.6.1 introduced two H-tensor synthesis equations. The first model, (3.35), uses the full wideband correlation matrix $\mathbf{R}_{\mathrm{WB},H}$ to generate an H-tensor and thus needs $(M_{\mathrm{Rx}}M_{\mathrm{Tx}}D)^2$ parameters. The second model, (3.36), uses the eigenvalues and eigenvectors of $\mathbf{R}_{\mathrm{WB},H}$ to generate an ensemble of H-tensors. Because of the eigenvalues, this model requires $M_{\mathrm{Rx}}M_{\mathrm{Tx}}D + (M_{\mathrm{Rx}}M_{\mathrm{Tx}}D)^2$ parameters. Section 3.4.2 showed that the Kronecker model (3.19) requires $D(M_{\mathrm{Rx}}^2 + M_{\mathrm{Tx}}^2)$ parameters.

The structured model (3.57) reduces the number of parameters by expressing the eigenvectors of the correlation matrix as the Kronecker product of three smaller

Table 3.1 Total number of parameters required by each synthesis equation, with a typical example

Synthesis Equation Number	Model Description	Number of Required Parameters	Number of Parameters for $(M_{Rx}, M_{Tx}, D) = (4, 4, 4)$
(3.35)	Wideband, full correlation tensor	$(M_{Rx}M_{Tx}D)^2$	4096
(3.36)	Wideband, EVD of correlation tensor	$M_{Rx}M_{Tx}D + (M_{Rx}M_{Tx}D)^2$	4160
(3.19)	Kronecker model	$D(M_{Rx}^2 + M_{Tx}^2)$	128
(3.57)	Structured model	$M_{Rx}M_{Tx}D + (M_{Rx}^2 + M_{Tx}^2 + D^2)$	112

vectors and thus only requires $M_{Rx}M_{Tx}D + (M_{Rx}^2 + M_{Tx}^2 + D^2)$ parameters to synthesize an exemplar H-tensor. Table 3.1 summarizes the number of parameters needed for each synthesis equation, along with the example when $M_{Rx} = 4$, $M_{Tx} = 4$, and $D = 4$.

3.7 SUMMARY AND DISCUSSION

3.7.1 The Kronecker Model

The Kronecker model begins with the assumption that scatterers at both link-ends can be treated independently. Given this separability assumption, the model approximates the full correlation as the Kronecker product of one-sided correlation matrices. Hence, the number of parameters required is reduced to describe the channel and introduces some structure in the model. The one-sided correlation matrices are a fingerprint of the scattering seen at either link-end. The wideband-equivalent synthesis equation is given as

$$\mathbf{H}_{\text{Kron}}[d] = \mathbf{R}_{Rx}^{1/2}[d]\mathbf{G}[d](\mathbf{R}_{Tx}^{1/2}[d])^T. \tag{3.58}$$

Despite its popularity, the Kronecker model, and the separability assumption specifically, has been shown to be inaccurate. The separability assumption leads to artifact paths in the MIMO APS. Using real-life data, in Chapter 6, we show the Kronecker model consistently underestimates the capacity of the channel.

3.7.2 The Weichselberger Model

The Weichselberger model is a narrowband model designed specifically to address the shortcomings of the separability assumption. It assumes that scatterers at either link-end are coupled and quantifies this as the average power coupled between eigenvectors of the channel correlation. The Weichselberger synthesis equation is thus defined by

$$\mathbf{H}_{\text{Weich}} = \mathbf{U}_{Rx}(\tilde{\mathbf{\Omega}} \odot \mathbf{G})\mathbf{U}_{Tx}^T. \tag{3.59}$$

The coupling matrix $\boldsymbol{\Omega}$ reflects the spatial structure of the channel. The Weichselberger model has been shown in the literature to be more accurate than the Kronecker model when estimating both the capacity and the spatial structure of the channel.

3.7.3 The Structured Model

The structured model is the extension of the Weichselberger model to the wideband MIMO channel. It uses tensor calculus to describe the coupling between scatterers across receive-transmit-delay space. In deriving the structured model, we introduced several concepts.

Tensor calculus was used to define three new correlation matrices, which are collectively termed the one-sided correlation matrices $\mathbf{R}_{\text{Rx}}$, $\mathbf{R}_{\text{Tx}}$, and $\mathbf{R}_{\text{Del}}$. These matrices described the correlation around the receiver, the transmitter, and in delay space, respectively. The wideband coupling coefficient ω_{mnd} is defined as the average energy coupled between the mth, nth, and dth eigenvector of $\mathbf{R}_{\text{Rx}}$, $\mathbf{R}_{\text{Tx}}$, and $\mathbf{R}_{\text{Del}}$, respectively.

After some analytic work, this led to the structured model synthesis equation,

$$\mathcal{H}_{\text{struct}} = \mathcal{G} \times_1 \mathbf{U}_{\text{Rx}} \times_2 \mathbf{U}_{\text{Tx}} \times_3 \mathbf{U}_{\text{Del}}. \tag{3.60}$$

The structured model approximates the correlation across receive-transmit-delay space. Unlike the Kronecker model, it does not assume independence between scatterers at the receiver, the transmitter, or at different delays. Because of this, the structured model better approximates the actual structure of the wideband MIMO channel. For most practical cases, the structured model also requires fewer parameters than the Kronecker model to synthesize an ensemble of H-tensors.

In Chapter 6, we use real-life data to verify the performance of the structured model. We compare its performance against that of the Kronecker model. Using several metrics, we show that the structured model outperforms the Kronecker model in almost all cases.

In the next chapter, we introduce another common type of MIMO channel model, *cluster models*. More recently, cluster models have focused on simulating the time-variant nature of mobile channels, a major shortcoming of correlative models. Cluster models are also closer to ray-tracing models and are thus much more complex than correlative models. They have been shown, however, to be fairly accurate, and form the basis for most standardized models.

3.8 NOTES AND REFERENCES

3.8.1 Correlative Models

Kronecker model, and the concept of one-sided correlation, arose from discussions contained in Refs. 31 and 35. Pedersen et al. (72), followed by Yu et al. (76), were the first to extend the Kronecker model to the wideband case.

In Ref. 51, Weichselberger presents the vector and matrix synthesis equations and provides a detailed analysis of the Weichselberger model.

The structured model was first presented in Ref. 77.

3.8.2 Tensor Decomposition

Concepts from the HOSVD were used in the derivation of the structured model. The HOSVD is an extension of the matrix SVD to higher-order tensors. Its derivation, properties, and some practical applications are discussed in Ref. 75. Tensors and tensor decomposition methods such as the HOSVD are powerful tools in that they can be extended to higher dimensions. For example, none of the correlative models above include the effects of time variation. Adding another dimension onto the H-tensor to include time makes it possible to analyze the time-variation of the wideband MIMO channel and its relation to delay (similar to Bello's model).

4

CLUSTER MODELS

One of the major drawbacks of many correlative channel models is their inability to model time-variation. Some channel statistics, such as received power, can change dramatically with small variations in transmitter/receiver position. Thus, it is important, especially in a mobile environment, to model the time-variant nature of the channel.

Cluster models bridge the gap between correlative models, which are largely stochastic in nature, and ray-tracing models, which rely more heavily on a deterministic geometry. Physically, clusters are associated with groups of scatterers in the channel. The energy reflected from these scatterers is located at different points in space. This leads to energy arriving at different *angles of arrival* (AoAs) and *angles of departure* (AoDs), as viewed from the receiver or transmitter, respectively (78). The different path lengths also lead to different *delays*. The size of the cluster is determined by its *spread*, as seen from both link-ends. The cluster can be broken down into a number of *multipath components*, each with its own AoA, AoD, and power. Cluster modeling reduces the behavior of clusters to a set of stochastic parameters. For most cluster models, the parameters are measured from real environments. Thus, cluster models reflect real-life channels in that their parameters are taken from measurements.

More recently, cluster models have focused on time-variations due to movement. The movement may be at the receiver, transmitter, or in the clusters themselves. Even though most cluster parameters are stochastic, cluster models are still largely geometry-based. This means that any movement can be tracked, and the received signal can be recomputed given a new geometry. Thus, cluster models become an intuitively satisfying way of simulating movement within the channel.

Multiple-Input, Multiple-Output Channel Models: Theory and Practice. By Nelson Costa and Simon Haykin
Copyright © 2010 John Wiley & Sons, Inc.

In this chapter, we begin with a discussion of clusters. We then discuss how clusters are identified. We then move on to describe several cluster models that occur frequently in the literature. The first is the *Saleh–Valenzuela model*, which describes how clusters behave in delay space. The *extended Saleh–Valenzuela model* extends the concept of clustering to the azimuth domain. The *European Cooperation in the field of Scientific and Technical Research* (COST) initiative has produced a number of channel models that form the basis for other standardized models used today, such as the *3rd Generation Partnership Project spatial channel model* (3GPP SCM) (34) and the *Wireless World Initiative New Radio II* (WINNER II) channel model (79). The *COST 273 model* is one of the more popular and one of the most comprehensive models to come from the COST initiative; we cover its implementation in Section 4.5. The *random cluster model* (RCM) extends the COST 273 model and focuses on the time-variant nature of the channel. The RCM includes a framework for automatically identifying and tracking clusters from measured data, allowing for their movement to be accurately modeled. We summarize the RCM and major contributions in Section 4.6.

4.1 WHAT IS A CLUSTER?

It turns out that there is no universally agreed upon definition of a cluster. From the discussion in Chapter 2, energy can take many paths from the transmitter to the receiver. At the highest level, energy at the receiver arrives in groups, loosely referred to as clusters. Clusters can be attributed to the discrete nature of scatterers in a physical channel; each scatterer will form a path, or a number of paths, that carries energy from the transmitter to the receiver. At the receiver, these clusters arrive at different times and from different angles. Therefore, clusters are evident in the delay domain, as well as higher-order angular-delay domains.

Each cluster consists of a number of multipath components. At the highest level, a cluster can be thought of as a group of multipath components having similar parameters. However, one's definition of a cluster depends largely on how the multipath components are grouped together, or how "similar" the multipath components are. Identifying clusters can be done *visually* by way of the APS or *automatically* using *clustering algorithms*. Visually identifying clusters leads to the most subjective definition of a cluster, depending on the space in which clusters are being identified (delay, double-directional angular, etc.). Identifying all clusters in real-life data is often difficult, if not impossible. In these instances, it is unlikely that two people will identify the same clusters given the same data set. In addition, identifying clusters in large data sets is cumbersome (80).

Until recently, clusters were most often identified visually. It was Turin et al. (81) who were among the first to recognize that energy arrived at the receiver in clusters and that the clusters followed a Poisson process. Saleh and Valenzuela (82), building on the work of Turin et al., showed that energy in a wideband channel arrived in clusters and that each cluster consisted of several resolvable multipath components (Fig. 4.1). The concept of clusters was used to explain the multiple paths that manifested in the

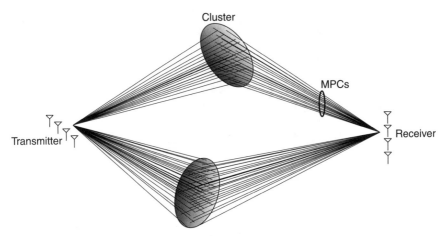

Figure 4.1 A cluster-based channel. Clusters are decomposed into a number of resolvable multipath components and can be viewed differently at either link-end. MPCs, multipath components.

delay domain. The concept of cluster-based returns was extended to the azimuth-delay domain in the COST 259 channel model (83, 84). Laurila et al. identified clusters in the azimuth-delay domain (85). Chong et al. used the single-directional channel model and the concept of clustering to develop a statistical wideband channel model (86). Using measurement data, Yu et al. measured cluster angular spreads in indoor wideband channels for the 802.11 TGn standard (87). The first to identify clusters in the double-directional azimuth-delay domain was Czink et al. (88). This was an improvement on cluster identification as clusters tend to be more concentrated in the AoA–AoD domain and harder to identify across different delays.

More recently, several automatic clustering algorithms have been introduced. Using automatic clustering algorithms presents a repeatable and automatic method of identifying clusters. Each automatic clustering algorithm introduces its own definition of a "cluster," but often has the advantage of being repeatable. Salo et al. (89) developed a semiautomatic method that identified clusters in the azimuth and delay domains separately, using a hierarchical clustering algorithm. In this way, the method did not estimate clusters jointly across domains. After being identified in each domain, the clusters had to be associated visually across them. In Ref. 80, Czink describes the *KPowerMeans* clustering algorithm. The algorithm jointly identifies clusters across the AoA–AoD–delay domains. In addition, both azimuth and elevation are considered.

4.2 THE SALEH–VALENZUELA MODEL

The most basic description of clusters is in the delay domain. In 1972, while performing wideband measurements for the development of vehicle location systems, Turin et al. (81) were the first to observe that the arrival times for energy from different

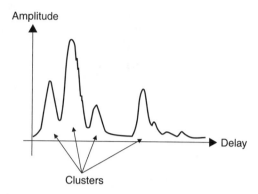

Figure 4.2 An example PDP showing cluster arrivals in the delay domain.

paths followed a single Poisson distributed process. Figure 4.2 shows a typical power delay profile (PDP) at the receiver, indicating individual clusters. Although their model could explain the majority of their observations, it could not account for the initial paths in the PDP, which tended to arrive in "groups."

The seminal paper by Saleh and Valenzuela (82) presents a model that describes the arrival times of multipath components in the delay domain for a time-invariant channel. The model is relatively simple. Over the years, the model has been shown to be quite accurate. Elements of the model can be seen in many channel models published since then, including the 3GPP *spatial channel model* (SCM) (90), COST 259 (83), COST 273 (91), and the 802.15 *ultra wide band* (UWB) (92) models.

Some of the more notable contributions of the Saleh–Valenzuela model are the following:

- It was one of the first models that considered the clustering of multipath components.
- The model divided the received PDP into the composition of a number of cluster PDPs and an overall PDP. Both PDPs were modeled as exponentially decaying.
- The arrival times of the clusters and the multipath components within each cluster were modeled as two independent Poisson processes.

The Saleh–Valenzuela model builds on the work of Turin et al. by breaking the channel response into clusters, each consisting of a number of multipath components. Both the arrival times of the clusters and of the multipath components within the cluster were modeled as independent Poisson processes, each with a different rate. Both the overall PDP and the PDP of each cluster were modeled as exponentially decaying. Furthermore, the slope of the cluster PDP was typically steeper than the overall PDP, as shown in Fig. 4.3.

Herein, we summarize the Saleh–Valenzuela model and discuss its implementation. In Section 4.4, we describe a simple extension of the Saleh–Valenzuela model to the azimuth domain.

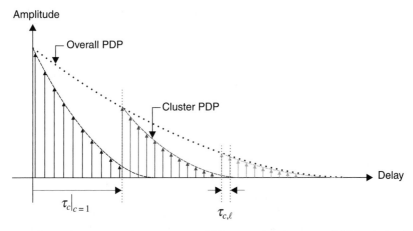

Figure 4.3 The Saleh–Valenzuela model PDP, denoting the overall PDP, cluster PDP, cluster arrival times, and multipath component arrival times.

4.2.1 Model Summary

The Saleh–Valenzuela model begins with the assumption that multipath components arrive in clusters. The arrival time of each cluster is defined as the arrival of the first multipath component associated with that cluster. The *cluster arrival time* τ_c is a Poisson process with rate λ_c, whose conditional distribution is given by

$$p(\tau_c|\tau_{c-1}) = \lambda_c e^{-\lambda_c(\tau_c - \tau_{c-1})} \quad \text{for } c > 0, \tag{4.1}$$

where $\tau_0 = 0$ is assumed. Let $\tau_{c,\ell}$ be the *arrival time* of the ℓth multipath component in cluster c. The multipath component arrival time is also a Poisson process, with rate $\lambda_{c,\ell}$, and is independent of τ_c. The conditional distribution is given by

$$p(\tau_{c,\ell}|\tau_{c,\ell-1}) = \lambda_{c,\ell} e^{-\lambda_{c,\ell}(\tau_{c,\ell} - \tau_{c,\ell-1})} \quad \text{for } \ell > 0, \tag{4.2}$$

where the delay of the first multipath component in each cluster is assumed to be zero (i.e., $\lambda_{c,0} = 0$).

The time-invariant channel impulse response $h(\tau)$ is computed as the sum contribution of all multipath components in all clusters,

$$h(\tau) = \sum_{c=0}^{\infty} \sum_{\ell=0}^{\infty} a_{c,\ell} e^{j\varphi_{c,\ell}} \delta(\tau - \tau_c - \tau_{c,\ell}), \tag{4.3}$$

where $a_{c,\ell}$ and $\varphi_{c,\ell}$ are the *gain* and the *phase* of the ℓth multipath component in cluster c, respectively. The phase term depends on the overall path length and changes rapidly for short variations in the receiver position. Thus, $\varphi_{c,\ell}$ is assumed to be uniformly distributed on $[0, 2\pi)$. The gain term $a_{c,\ell}$ is real and governs the PDP of each cluster, as well as the slope of the overall PDP. Its mean squared value $\overline{a_{c,\ell}^2}$ can be approximated as a function of two decaying exponentials, as shown by

$$\overline{a_{c,\ell}^2} = \overline{a_{0,0}^2} e^{-\tau_c/\gamma_c} e^{-\tau_{c,\ell}/\gamma_{c,\ell}}, \tag{4.4}$$

where $\overline{a_{0,0}^2}$ is the PDP initial amplitude, and γ_c, $\gamma_{c,\ell}$ are the cluster and multipath component *PDP time constants*, respectively. These constants are important; they are a direct measure of the severity of multipath fading in a channel, or, equivalently, a measure of the diversity. They are also directly related to the *delay spread*.

The gain of the first multipath component in the first cluster, $\overline{a_{0,0}^2}$, can be set arbitrarily to scale the PDP. Saleh and Valenzuela choose this value to reflect the average power received when the receiver and transmitter are separated by 1 m, and use the approximation

$$\overline{a_{0,0}^2} \cong \frac{1}{\gamma_{c,\ell}\lambda_{c,\ell}} G_{\text{Rx}} G_{\text{Tx}} \frac{\lambda_0^2}{(4\pi)^2} r^{-(1+\alpha)} \Big|_{r=1}, \tag{4.5}$$

where G_{Rx}, G_{Tx} are the receive-antenna and transmit-antenna gains, $\lambda_0 = 2\pi f_c$, and α is the *pathloss exponent*, which is an empirical value.

The gains themselves are Rayleigh distributed random variables, with probability density function (PDF)

$$p(a_{c,\ell}) = \frac{2a_{c,\ell}}{\overline{a_{c,\ell}^2}} e^{a_{c,\ell}^2/\overline{a_{c,\ell}^2}}. \tag{4.6}$$

In general, clusters can overlap; that is, given some c, it is possible to have a $\tau_{c,\ell}$ that satisfies $\tau_{c,\ell} > \tau_{c+1} - \tau_c$ where the time period between clusters is given by $\tau_{c+1} - \tau_c$. In practice, the clusters decay quicker than the overall PDP (i.e., $\gamma_c > \gamma_{c,\ell}$). Even though the summations in (4.3) involve an infinite sum, the multipath components for each cluster tend to disappear below the noise floor, oftentimes before the arrival of the next cluster. The result is that clusters often appear disjoint and thus can be distinguished visually.

The number of additional clusters $n_{c,\text{add}}$ (in addition to the shortest path cluster for which $c = 0$) is modeled by the Poisson distribution

$$p[n_{c,\text{add}}] = \frac{(T_o\lambda_{c,\text{add}})^{n_{c,\text{add}}}}{n_{c,\text{add}}!} e^{-T_o\lambda_{c,\text{add}}}, \tag{4.7}$$

where T_o is the observation period, and $\lambda_{c,\text{add}}$ is the mean, which can be computed from the observed data. The total number of clusters is thus given as $N_c = n_{c,\text{add}} + 1$.

Table 4.1 Measured parameters for the Saleh–Valenzuela model in an indoor, fixed point-to-point office environment (82).

Parameter	Description	Measured Value
γ_c	Cluster PDP time constant	60 ns
$\gamma_{c,\ell}$	Multipath component PDP time constant	20 ns
$1/\lambda_c$	Cluster arrival rate	$\approx$300 ns
$1/\lambda_{c,\ell}$	Multipath component arrival rate	$\approx$5 ns
α	Free space pathloss exponent	Between 3 and 4

4.2.2 Model Implementation

Given the model description above, the steps required to generating an exemplar $h(\tau)$ is relatively straightforward. The following is a point-form summary of the procedure given in Ref. 82. The procedure assumes that certain parameters have been computed from the measured data.

- Randomly generate the cluster arrival times τ_c using (4.1) and given the rate λ_c. Set $\tau_0 = 0$ (arbitrary).
- Estimate $\overline{a_{0,0}^2}$ from (4.5), given the parameters G_{Rx}, G_{Tx}, $\gamma_{c,\ell}$, $\lambda_{c,\ell}$, and α. Alternately, the value of $\overline{a_{0,0}^2}$ can be chosen to arbitrarily scale the data (and possibly remove the absolute power reference).
- Generate the multipath component arrival times for each cluster $\tau_{c,\ell}$ using (4.2) and given $\lambda_{c,\ell}$.
- Generate a set of path gains $a_{c,\ell}$ using (4.4) and (4.6) given the parameters $\overline{a_{0,0}^2}$, time constants γ_c, $\gamma_{c,\ell}$, and the gain $\overline{a_{0,0}^2}$.
- Finally, the phases are drawn randomly according to $\varphi_{c,l} \sim U[0, 2\pi)$.

4.2.3 Some Typical Parameters

Saleh and Valenzuela performed an extensive measurement campaign from many fixed indoor point-to-point locations at Bell Labs (82). They were able to estimate the parameters heuristically using curve fitting techniques. Table 4.1 summarizes the results.

4.3 CLUSTERS IN TIME AND SPACE

Since the Saleh–Valenzuela model, there have been numerous cluster models in the literature that extend the description of clusters to include their location in space. A cluster location is most often denoted using a spherical coordinate system relative to the receive or transmit array. Thus, the position of each cluster c is given by the triplet $(\varphi_{X,c}, \theta_{X,c}, d_{X,c})$, where $\varphi_{X,c}$ and $\theta_{X,c}$ are the azimuth and elevation angles, $d_{X,c}$ is the

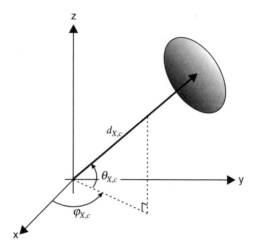

Figure 4.4 The symbol convention for cluster azimuth, elevation, and distance in relation to the Cartesian axes is illustrated.

distance of the cth cluster, and $X \in \{Rx, Tx\}$ (Fig. 4.4). The coordinates usually refer to the center of the cluster. It is important to note that the cluster position can be defined relative to the receiver and/or the transmitter. This leads to a *double-directional* description of the channel, which we discuss further in Section 4.5.1.

Conventionally, when viewing the channel from the receiver, $(\varphi_{Rx,c}, \theta_{Rx,c})$ refers to the AoA, whereas $(\varphi_{Tx,c}, \theta_{Tx,c})$ refers to the AoD. In the case of a *single-bounce cluster*, The total *path distance* $d_{Rx,c} + d_{Tx,c}$ is directly related to the cluster delay τ_c,

$$\frac{d_{Rx,c} + d_{Tx,c}}{c_0} \leq \tau_c. \tag{4.8}$$

In practice, because of the difficulty in simultaneously obtaining accurate azimuth and elevation measurements, and because the majority of energy arriving at the receiver in most environments is concentrated in the azimuth plane, many directional descriptions of the channel omit elevation parameters.

4.3.1 Azimuth, Elevation, and Delay Spreads

Clusters are composed of a number of multipath components. Each multipath component can be thought of as a discrete contribution to the cluster, randomly distributed within the cluster in all three domains (azimuth, elevation, and delay). Most cluster models hold that multipath components are randomly distributed in each domain independently, according to some distribution. Each distribution is often characterized by its second central moment. This parameter is most commonly referred to as the cluster *receiver spread*, or simply *spread*. The cluster spread is a measure of the cluster dispersion in each domain.

4.3.1.1 *Cluster Delay Spread*

One of the oldest statistics of a wideband channel, the *receiver delay spread* (41), first defined in Section 2.2.3, is defined as the second central moment of the PDP $P(\tau)$,

$$\sigma_\tau^2 = \mathrm{E}\{\tau^2\} - \mathrm{E}^2\{\tau\}, \tag{4.9}$$

where

$$\mathrm{E}\{\overline{\tau^n}\} = \frac{\int \tau^n P(\tau)\, d\tau}{\int P(\tau)\, d\tau}. \tag{4.10}$$

In cluster modeling, the discrete form of the above equations is commonly used. Also, the delay spread of each cluster is computed individually. The expectation operator may also be approximated using a statistical average. Consider the case where the cth cluster consists of $\ell = \{1, \ldots, L_c\}$ paths. At time interval i, each path has a *complex* amplitude $A_{c,\ell}[i]$ and is associated with delay $\tau_{c,\ell}[i]$. The delay is directly proportional to the path distance. The receiver delay spread for cluster c can be estimated using (93),

$$\sigma_{\tau,c} = \sqrt{\frac{\sum_{i=1}^{T} \sum_{\ell=1}^{L_c} (\tau_{c,\ell}[i] - \overline{\tau}_c)^2 |A_{c,\ell}[i]|^2}{\sum_{i=1}^{T} \sum_{\ell=1}^{L_c} |A_{c,\ell}[i]|^2}}, \tag{4.11}$$

where the mean cluster delay $\overline{\tau}_c$ is defined as

$$\overline{\tau}_c = \frac{\sum_{i=1}^{T} \sum_{\ell=1}^{L_c} \tau_{c,\ell}[i] |A_{c,\ell}[i]|^2}{\sum_{i=1}^{T} \sum_{\ell=1}^{L_c} |A_{c,\ell}[i]|^2}. \tag{4.12}$$

4.3.1.2 *Azimuth and Elevation Spread Estimation*

The cluster spread in the azimuth and elevation domains is dependent on the number of multipath components in a given cluster, their power, and their direction. Similar to the delay spread, the azimuth power spread $\sigma_{\varphi,c}$ can be estimated by

$$\sigma_{\varphi,c} = \sqrt{\frac{\sum_{i=1}^{T} \sum_{\ell=1}^{L_c} (\varphi_{c,\ell}[i] - \overline{\varphi}_c)^2 |A_{c,\ell}[i]|^2}{\sum_{i=1}^{T} \sum_{\ell=1}^{L_c} |A_{c,\ell}[i]|^2}}, \tag{4.13}$$

where the mean cluster azimuth angle $\overline{\varphi}_c$ is defined as

$$\overline{\varphi}_c = \frac{\sum_{i=1}^{T} \sum_{\ell=1}^{L_c} \varphi_{c,\ell}[i] |A_{c,\ell}[i]|^2}{\sum_{i=1}^{T} \sum_{\ell=1}^{L_c} |A_{c,\ell}[i]|^2}. \tag{4.14}$$

By replacing φ with the elevation angle θ in (4.13) and (4.14), we can compute the *elevation spread* $\sigma_{\theta,c}$. The above definitions for azimuth and elevation spread can also be extended to the double-directional channel (Section 4.5.1), where the channel is resolved at both link-ends. In the following, we denote the AoA spreads in the azimuth and elevation as $\sigma_{Rx,\varphi}$ and $\sigma_{Rx,\theta}$, respectively. Similarly, we denote the AoD spreads in the azimuth and elevation as $\sigma_{Tx,\varphi}$ and $\sigma_{Tx,\theta}$, respectively.

4.4 THE EXTENDED SALEH–VALENZUELA MODEL

The sounder used by Saleh and Valenzuela during their experiments employed antennas that were omnidirectional in the azimuth plane. Thus, the Saleh–Valenzuela model omits any AoA, AoD characterization. Numerous publications since then have extended the model to include the azimuth and elevation domains. Using a sounder with a positionable parabolic antenna at the receiver, Spencer et al. (94) measured PDPs in the azimuth domain, and extended the Saleh–Valenzuela model to include the azimuth AoA. The resulting model, like the Saleh–Valenzuela model, is only applicable to the time-invariant channel. One important finding was that, in addition to being clustered in the delay domain, multipath components were observed to be clustered in the azimuth domain. The azimuth clustering followed a Laplacian distribution, as we discuss in the following.

The extended Saleh–Valenzuela model describes the channel impulse response in azimuth-elevation space as the sum of the contribution from all multipath components in all clusters,

$$h(\tau, \varphi) = \sum_{c=0}^{\infty} \sum_{\ell=0}^{\infty} a_{c,l} e^{j\varphi_{c,l}} \delta(\tau - \tau_c - \tau_{c,l}) \delta(\varphi - \varphi_c - \varphi_{c,l}), \qquad (4.15)$$

where the variables φ_c and $\varphi_{c,l}$ describe the azimuth AoA of the clusters and multipath components, respectively. Here, φ_c represents the *mean value* of the AoA of all multipath components within that cluster.

Both φ_c and $\varphi_{c,l}$ are random variables. Given the AoA of the first cluster φ_0, the condition distribution $p(\varphi_c | \varphi_0)$ is modeled as being U[0, 2π). The multipath component distribution is modeled as a zero-mean Laplacian:

$$p(\varphi_{c,\ell}) = \frac{1}{\sqrt{2}\sigma_{\varphi c,\ell}} e^{\frac{-\sqrt{2}}{\sigma_{\varphi c,\ell}}|\varphi_{c,\ell}|}, \qquad (4.16)$$

where the variance $\sigma_{\varphi c,\ell}^2$ is the *azimuth spread* at the receiver, which is an empirical quantity. In a manner similar to the time constants γ_c and $\gamma_{c,\ell}$, the variance $\sigma_{\varphi c,\ell}^2$ quantifies the multipath diversity, but in *space*. This has an intuitive physical interpretation. For example, channels with many large scatterers arranged with respect to increasing range, or many strong multibounce paths of varying lengths, will often lead to large delay spread. Channels with many scatterers at the same range and clustered together, or very large objects (relative to the field-of-view) such as buildings, often lead to large azimuth spreads.

4.5 THE COST 273 MODEL

Through their pioneering work, Saleh and Valenzuela were able to show that multipath components were clustered in the delay domain. Spencer et al. were among the first to show, using measured data, that multipath components could be clustered in the azimuth as well as delay domains. Since that time, more sophisticated sounders have been built, having the ability to measure the AoA and AoD of multipath components at both the receiver and the transmitter. This has facilitated the development of models that include the effects of clustering in the azimuth, elevation, and delay domains. Measuring these parameters at both link-ends leads to the so-called *double-directional* description of the channel (78). In parallel, super-resolution algorithms have been developed, allowing for the automated detection of multipath components within measured data. Some examples are the *space-alternating generalized expectation maximization* (SAGE) algorithm (95) and its variant *initialization and searching improved space-alternating generalized expectation maximization* (ISIS) (96). These algorithms are similar to the beamforming technique first discussed in Section 2.6.1, but with increased resolution. In this way, they are used to resolve AoDs of individual multipath components in dense multipath data, both in angle and delay.

More recently, cluster models have begun to focus on the time-variant nature of the channel, especially in cases where the receiver and/or transmitter are mobile. To model large-scale fading, these models include some mechanism to simulate cluster movement, and/or some process by which clusters fade in and out of view. In addition to this, the models have been generalized to include the MIMO channel.

The COST 273 model (91), a refinement of the COST 259 model (83), is the result of a large coordinated effort in the European scientific and industrial communities to develop a comprehensive channel model that can, in turn, be used in the development of radio communication standards. The final report (91) summarizes all major findings, including measurement campaigns and results, signal processing techniques, and some design trade-offs. The report also includes a summary of the COST 273 model. The model incorporates findings from the COST 273 and 259 initiatives, as well as other existing models.

Part of the work that resulted from the COST 273 initiative was a MATLAB implementation of the model for a *large urban macrocell* environment. The scripts are available online (97). The discussion here closely follows the implemented version, rather than the full framework presented in the final report by Correia et al. (91). The MATLAB implementation does not include the following features: visibility regions, diffuse radiation, autocorrelation distances, and cross-polarization effects. The differences between the implemented version and the complete framework are discussed further in Section 4.5.5.

4.5.1 Generic Channel Model

The *generic channel model* (GCM) was first presented in Ref. 98. The model derives its name from the fact that it is *generic* enough to be able to model many different types of channels. The COST 273 channel model uses the same GCM for all environments.

By itself, the GCM is a method of computing the double-directional impulse response given the multipath component statistics, as well as some external parameters.

At its highest level, the GCM consists of two main equations. The first describes the contribution of the multipath components to the impulse response of the channel, called the *double-directional impulse response* (78),

$$h(\tau, \varphi_{Rx}, \theta_{Rx}, \varphi_{Tx}, \theta_{Tx}) = \sum_{\ell=1}^{L} h_\ell(\tau, \varphi_{Rx,\ell}, \theta_{Rx,\ell}, \varphi_{Tx,\ell}, \theta_{Tx,\ell}), \qquad (4.17)$$

where τ is the delay, $\varphi_{Rx,\ell}$ and $\theta_{Rx,\ell}$ are the azimuth and elevation AoA of the ℓth path, $\varphi_{Tx,\ell}$ and $\theta_{Tx,\ell}$ are the azimuth and elevation AoD of the ℓth path, and L is the total number of multipath components. The contribution of each multipath component is explicitly given by

$$h_l(\tau, \varphi_{Rx}, \theta_{Rx}, \varphi_{Tx}, \theta_{Tx}) = a_\ell e^{j\varphi_\ell} \delta(\tau - \tau_\ell) \delta(\varphi_{Rx} - \varphi_{Rx,\ell})$$
$$\times \delta(\theta_{Rx} - \theta_{Rx,\ell}) \delta(\varphi_{Tx} - \varphi_{Tx,\ell}) \delta(\theta_{Tx} - \theta_{Tx,\ell}) \qquad (4.18)$$

where each multipath component has a complex amplitude $A_\ell = a_\ell e^{j\varphi_\ell}$, delay τ_ℓ, and $\phi_\ell = j2\pi\tau_\ell f_c$.

The double-directional impulse response characterizes the contribution, or path weight, of each multipath component without system considerations, such as number of antennas, antenna patterns, antenna locations at each array, and the center frequency. These parameters often remain constant regardless of the channel. Thus, the double-directional impulse response characterizes the channel itself, before considering a transmitter/receiver configuration.

Given the double-directional impulse response, the $M_{Rx} \times M_{Tx}$ elements of the H-matrix are computed as

$$h_{mn} = \sum_{\ell=1}^{L} h_\ell(\tau, \varphi_{Rx,\ell}, \theta_{Rx,\ell}, \varphi_{Tx,\ell}, \theta_{Tx,\ell})$$
$$\times G_{Rx}(\varphi_{Rx,\ell}, \theta_{Rx,\ell}) G_{Tx}(\varphi_{Tx,\ell}, \theta_{Tx,\ell})$$
$$\times \exp(j\langle \mathbf{k}(\varphi_{Rx,\ell}, \theta_{Rx,\ell}) \cdot \mathbf{x}_{Rx,m}\rangle) \exp(j\langle \mathbf{k}(\varphi_{Tx,\ell}, \theta_{Tx,\ell}) \cdot \mathbf{x}_{Tx,n}\rangle) \qquad (4.19)$$

where $G_{Rx}(\varphi_{Rx,\ell}, \theta_{Rx,\ell})$ and $G_{Tx}(\varphi_{Tx,\ell}, \theta_{Tx,\ell})$ are the antenna patterns at the receiver and transmitter, respectively, and

$$\langle \mathbf{k}(\varphi, \theta) \cdot \mathbf{x} \rangle = \frac{2\pi}{\lambda}(x \cos\theta \cos\varphi + y \cos\theta \sin\varphi + z \sin\theta) \qquad (4.20)$$

is the inner product of the wave vector $\mathbf{k}(\varphi, \theta)$ with an antenna at position $\mathbf{x} = [x \ y \ z]$, relative to the center of the array. Because (4.19) models the effects of the measurement system, it is sometimes referred to as the *system model*.

The GCM defines how the multipath components add to form the channel impulse response, but it does not tell us how to compute the gain and phase of each multipath component. It does not assume any channel geometry or impose any statistics on AoAs, AoDs, and so forth. It does not even assume that multipath components arrive in clusters. The GCM simply expresses the channel as the sum contribution of multipath components. As we will see, the COST 273 framework focuses on a statistical model for the amplitude, phase, AoA, AoD, and delay of all multipath components. The parameters of the statistical model change according to the *environment*. In addition, the COST 273 model extends the GCM to the case where multipath components can be clustered.

4.5.2 Environments

The COST 273 is defined differently for different environments. The set of all possible channels is divided into four *environments*, each having a number of *propagation scenarios*. The COST 273 final report (91) lists parameters for one propagation scenario in each of three environments:

- Macrocell: Large urban macrocell
- Microcell: Urban center
- Picocell: Halls

Depending on the environment, both the receiver and the transmitter are allowed to move. The COST 273 environments are largely differentiated by the receiver and transmitter location and the density of the surrounding buildings. A different set of parameters is given for each environment. The set of all parameters is divided into *external parameters* that are user defined and remain fixed for a simulation run and *stochastic parameters* that provide the information necessary to generate all random variables in the model. Most of these were compiled from numerous measurement campaigns and thus are designed to reflect the *average* environment.

4.5.3 Receiver, Transmitter Placement

Many models designed specifically for cellular channels often refer to one link-end as the *base station* (BS) and the other as the *mobile station* (MS). To be consistent with the notation in the rest of the book, we denote the BS as the transmitter and the MS as the receiver.

The first step in simulating any environment is to place the receiver and transmitter. For all environments, the transmitter is placed at the center of the cell and is described by the vector $\mathbf{r}_{Tx} = [0 \ \ 0 \ \ h_{Tx}]$, where h_{Tx} is the height of the transmitter array. The receiver is randomly placed within the cell at location $\mathbf{r}_{Rx} = [x_{Rx} \ \ y_{Rx} \ \ h_{Rx}]$. Time variability is introduced by moving the receiver, transmitter, or the clusters, depending on the environment. The receiver moves linearly with velocity $\mathbf{v}_{Rx}$. The transmitter moves linearly with velocity $\mathbf{v}_{Tx}$. For most environments, $\mathbf{v}_{Tx} = 0$ (i.e., the transmitter remains fixed).

To compute line-of-sight (LoS) statistics, the polar coordinates of the LoS path is given by $(d_{Rx,Tx}, \varphi_{Rx,Tx}, \theta_{Rx,Tx})$, where $d_{Rx,Tx}$ is the distance between the receiver and transmitter, $\varphi_{Rx,Tx}$ is the azimuth angle, and $\theta_{Rx,Tx}$ is the elevation angle. The *LoS delay* is computed as $\tau_0 = d_{Rx,Tx}/c_0$, where c_0 is the speed of light.

4.5.4 COST 273 Procedure

The COST 273 model includes three different types of clusters. *Single-interaction clusters* simulate the case when there is strong correlation between the AoA and AoD statistics. This can happen, for example, when both the receiver and the transmitter illuminate the same set of scatterers. *Local clusters* can be viewed as special instances of two single-interaction clusters. One cluster is centered at the receiver, and the other is optionally located at the transmitter. Local clusters aim to model scatterers at either link-end. *Twin clusters* simulate multiple-bounce scattering, where there is rich non–line-of-sight (NLoS) scattering between the receiver and transmitter.

4.5.4.1 Determining the Number and Type of Clusters   The total number of clusters N_c is governed by

$$N_c = N_{c.local} + N_{c.twin} + N_{c.single}, \tag{2.41}$$

where $N_{c.local}$, $N_{c.twin}$, and $N_{c.single}$ are the total number of local, twin, and single-interaction clusters, respectively. Not all cluster types are used for all environments. Although the model allows for local clusters around the transmitter and receiver, for all environments listed in Ref. 91, $N_{c,local} = 1$. For the macrocell scenario, the cluster is always at the receiver. The remaining cluster $N_{c,add} = N_{c,twin} + N_{c,single}$ is chosen such that

$$N_{c,add} = \max \begin{cases} 1 \\ X_p, \end{cases} \tag{4.22}$$

where X_p is a Poisson-distributed random variable with mean λ_p and PDF

$$p[k] = \frac{\lambda_p^k}{k!} e^{-\lambda_p}. \tag{4.23}$$

The parameter $K_{sel} \in [0, 1]$ determines the ratio of single-interaction to twin clusters, where $N_{c,twin} = \lceil K_{sel} N_{c,add} \rceil$.

4.5.4.2 Cluster Fading, Delay Spread, and Angular Spreads   The power of the cth cluster P_c is determined by the total cluster delay τ_c, and is given by

$$P_c = P_0 \cdot \max \begin{cases} e^{-k_\tau(\tau_c - \tau_0)} \\ e^{-k_\tau(\tau_b - \tau_0)} \end{cases}, \tag{4.24}$$

where k_τ is the cluster attenuation coefficient in dB/μs, τ_b is the cutoff delay, beyond which a cluster is considered insignificant, and τ_0 is the LoS delay.

Each cluster fades randomly, according to a log-normal distribution. This is equivalent to saying all multipath components within a given cluster are affected by the *fading factor* S_c. The delay and angular spreads of each cluster are also log-normally distributed and correlated with S_c, that is,

$$S_c = 10^{\sigma_s y_{s,c}/10}$$

$$\sigma_{\tau,c} = \mu_\tau 10^{\varepsilon \sigma_\tau y_{\tau,c}/10}$$

$$\sigma_{\varphi_{Rx},c} = \mu_{\varphi_{Rx}} 10^{\varepsilon \sigma_{\varphi_{Rx}} y_{\varphi_{Rx},c}/10}$$

$$\sigma_{\varphi_{Tx},c} = \mu_{\varphi_{Tx}} 10^{\varepsilon \sigma_{\varphi_{Tx}} y_{\varphi_{Tx},c}/10} \qquad (4.25)$$

$$\sigma_{\theta_{Rx},c} = \mu_{\theta_{Rx}} 10^{\varepsilon \sigma_{\theta_{Rx}} y_{\theta_{Rx},c}/10}$$

$$\sigma_{\theta_{Tx},c} = \mu_{\theta_{Tx}} 10^{\varepsilon \sigma_{\theta_{Tx}} y_{\theta_{Tx},c}/10},$$

where $\sigma_{\tau,c}$, $\sigma_{\varphi_X,c}$, $\sigma_{\theta_X,c}$ are the cluster spread in the delay, azimuth, and elevation domains, and $X \in \{Rx, Tx\}$. The parameter $\varepsilon = 0.5$ is assumed for all environments. The delay, azimuth, and elevation spreads are parameterized by (μ_τ, σ_τ), $(\mu_{\varphi,X}, \sigma_{\varphi,X})$, and $(\mu_{\theta,X}, \sigma_{\theta,X})$, all of which are specified for each environment. The parameters $y_{s,c}$, $y_{\tau,c}$, $y_{\varphi,c}$, and $y_{\theta,c}$ for every cluster c are correlated Gaussian random variables such that $y \sim N(0, 1)$. We describe how we generate these random variables in the following.

Reference 99 presents a method for generating correlated random variables using the Cholesky decomposition. The correlation coefficients $\rho_{x,y}$ for $x, y \in \{S_c, \sigma_{\tau,c}, \sigma_{\varphi_X,c}, \sigma_{\theta_X,c}\}$ are defined for each environment. The correlation coefficients can be arranged in a *correlation matrix* Γ, as shown by:

$$\Gamma = \begin{bmatrix} \rho_{S_c S_c} & \rho_{S_c \sigma_\tau} & \rho_{S_c \sigma_{\varphi Tx}} & \rho_{S_c \sigma_{\varphi Rx}} & \rho_{S_c \sigma_{\theta Tx}} & \rho_{S_c \sigma_{\theta Rx}} \\ \rho_{\sigma_\tau S_c} & \rho_{\sigma_\tau \sigma_\tau} & \rho_{\sigma_\tau \sigma_{\varphi Tx}} & \rho_{\sigma_\tau \sigma_{\varphi Rx}} & \rho_{\sigma_\tau \sigma_{\theta Tx}} & \rho_{\sigma_\tau \sigma_{\theta Rx}} \\ \rho_{\sigma_{\varphi Tx} S_c} & \rho_{\sigma_{\varphi Tx} \sigma_\tau} & \rho_{\sigma_{\varphi Tx} \sigma_{\varphi Tx}} & \rho_{\sigma_{\varphi Tx} \sigma_{\varphi Rx}} & \rho_{\sigma_{\varphi Tx} \sigma_{\theta Tx}} & \rho_{\sigma_{\varphi Tx} \sigma_{\theta Rx}} \\ \rho_{\sigma_{\varphi Rx} S_c} & \rho_{\sigma_{\varphi Rx} \sigma_\tau} & \rho_{\sigma_{\varphi Rx} \sigma_{\varphi Tx}} & \rho_{\sigma_{\varphi Rx} \sigma_{\varphi Rx}} & \rho_{\sigma_{\varphi Rx} \sigma_{\theta Tx}} & \rho_{\sigma_{\varphi Rx} \sigma_{\theta Rx}} \\ \rho_{\sigma_{\theta Tx} S_c} & \rho_{\sigma_{\theta Tx} \sigma_\tau} & \rho_{\sigma_{\theta Tx} \sigma_{\varphi Tx}} & \rho_{\sigma_{\theta Tx} \sigma_{\varphi Rx}} & \rho_{\sigma_{\theta Tx} \sigma_{\theta Tx}} & \rho_{\sigma_{\theta Tx} \sigma_{\theta Rx}} \\ \rho_{\sigma_{\theta Rx} S_c} & \rho_{\sigma_{\theta Rx} \sigma_\tau} & \rho_{\sigma_{\theta Rx} \sigma_{\varphi Tx}} & \rho_{\sigma_{\theta Rx} \sigma_{\varphi Rx}} & \rho_{\sigma_{\theta Rx} \sigma_{\theta Tx}} & \rho_{\sigma_{\theta Rx} \sigma_{\theta Rx}} \end{bmatrix}. \qquad (4.26)$$

To generate a set of correlated Gaussian random variables, we first take the Cholesky decomposition of Γ,

$$\Gamma = CC^T. \qquad (4.27)$$

Using a Gaussian random vector as an impetus, we generate a vector of correlated Gaussian random variables $\mathbf{y}$, yielding

$$\mathbf{y}_c = \mathbf{x}_c C, \qquad (4.28)$$

where the elements of $\mathbf{x}_c$, $x_i \sim N(0, 1)$. The structure of $\mathbf{y}$ is

$$\mathbf{y}_c = \left[y_{s,c}, y_{\sigma_s,c}, y_{\sigma_{\varphi BS},c}, y_{\sigma_{\varphi MS},c}, y_{\sigma_{\theta BS},c}, y_{\sigma_{\theta MS},c} \right] \tag{4.29}$$

where the elements of $\mathbf{y}$ are used in (4.25) to generate the correlated cluster statistics.

4.5.4.3 Cluster Position, Orientation, and Size   The following description applies to all cluster types. Cluster positions are described geometrically, using polar coordinates, as shown in Fig. 4.5a. Clusters are seen by both the receiver and transmitter. Thus, each cluster has two sets of coordinates, one each for the receiver, transmitter. The cluster AoA and AoD in the azimuth-elevation plane is described by $\varphi_{Rx,c}$, $\theta_{Rx,c}$ and $\varphi_{Tx,c}$, $\theta_{Tx,c}$, respectively. All clusters are circular in the azimuth-delay plane. The cluster diameter $d_{\tau,c}$ depends on the delay spread $\sigma_{\tau,c}$ such that

$$d_{\tau,c} = \frac{1}{2} \sigma_{\tau,c} c_0. \tag{4.30}$$

The total cluster delay τ_c is a function of the cluster delay seen by the transmitter and receiver and a *link delay* between clusters. Thus, τ_c is computed for each cluster as

$$\tau_c = \tau_{Rx,c} + \tau_{Tx,c} + \tau_{link,c}. \tag{4.31}$$

Local and single-interacting clusters can be considered as special cases of twin clusters, with $\tau_{link,c} = 0$.

Each cluster spans three axes, as shown in Fig. 4.5b, where the span $a_{X,c}$ is in delay, $b_{X,c}$ is in azimuth, and $h_{X,c}$ is in elevation. At the beginning of a simulation run, the

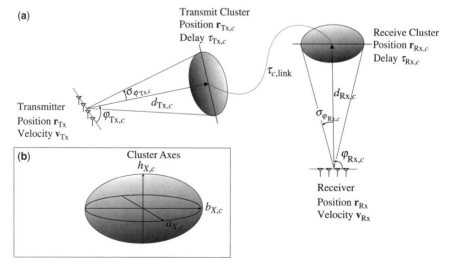

Figure 4.5   (a) Illustration of the symbol convention for transmit and receive clusters, including their relative locations, and (b) their spreads.

receive and the transmit clusters are rotated so that they "face" the receiver or the transmitter, respectively.

4.5.4.4 Local Clusters

Local clusters are centered at the transmitter, receiver, or both, depending on the environment. Thus, for a cluster at the receiver, $\mathbf{r}_{Rx,c} = \mathbf{r}_{Tx,c} = \mathbf{r}_{Rx}$. Similarly, for a cluster at the transmitter, $\mathbf{r}_{Rx,c} = \mathbf{r}_{Tx,c} = \mathbf{r}_{Tx}$. Next, we compute the spans of the cluster, $h_{X,c}$, $a_{X,c}$, and $b_{X,c}$ for $X \in \{Rx, Tx\}$. For the local clusters at the receiver or transmitter, the heights $h_{Rx,c}$ or $h_{Tx,c}$ are a function of the elevation spreads $\sigma_{\theta_{Tx},c}$ or $\sigma_{\theta_{Rx},c}$, as shown by

$$
\begin{aligned}
h_{Rx,c} &= d_{Rx,Tx} \tan(\sigma_{\theta_{Tx},c}) \\
h_{Tx,c} &= d_{Rx,Tx} \tan(\sigma_{\theta_{Rx},c}).
\end{aligned}
\tag{4.32}
$$

Note that the local cluster around the receiver is dependent on the elevation spread at the transmitter, and vice versa. The cluster is "circular" in the xy-plane. The remaining two axes are set equal to the cluster diameter (i.e., $a_{Rx,c} = b_{Rx,c} = d_{\tau,c}$). The cluster is "viewed" the same from the transmitter, so that $a_{Tx,c} = b_{Tx,c} = d_{\tau,c}$.

4.5.4.5 Twin Clusters

We begin by randomly placing each cluster in the azimuth-elevation-delay space. The AoA and AoD for each cluster is chosen from a set of marginal PDFs. These PDFs are defined for each environment in the COST 273 standard. For example, in the large urban macrocell implementation,

$$
\begin{aligned}
\varphi_{Rx,c}, \; \varphi_{Tx,c} &\sim U[0, 2\pi](\text{rad}) \\
\theta_{Rx,c}, \; \theta_{Tx,c} &\sim U[0, 1](\text{rad}).
\end{aligned}
\tag{4.33}
$$

The cluster delay τ_c is the same at the receiver, transmitter. For each cluster, τ_c is chosen from a marginal PDF, with one marginal PDF defined for each environment. In the large urban macrocell, τ_c is given by

$$
\tau_c = \tau_0 + X_\tau,
\tag{4.34}
$$

where τ_0 is the LoS delay, and $X_\tau \sim U[0, 0.5]$ (μs).

The cluster distances from the receiver, transmitter $d_{Rx,c}$, $d_{Tx,c}$ are computed as a function of the azimuth spreads $\sigma_{\varphi_{Rx},c}$, $\sigma_{\varphi_{Tx},c}$ as

$$
\begin{aligned}
d_{Rx,c} &= \frac{d_{\tau,c}}{2 \tan(\sigma_{\varphi_{Rx},c})} \\[2ex]
d_{Tx,c} &= \frac{d_{\tau,c}}{2 \tan(\sigma_{\varphi_{Tx},c})}.
\end{aligned}
\tag{4.35}
$$

Given τ_c, $d_{Rx,c}$, and $d_{Tx,c}$, the link delay $\tau_{link,c}$ is computed as

$$
\tau_{link,c} = \tau_c - \frac{d_{Rx,c} - d_{Tx,c}}{c}.
\tag{4.36}
$$

Next the cluster spans $h_{X,c}$, $a_{X,c}$, and $b_{X,c}$ are computed for $X \in \{Rx, Tx\}$. The heights of the receiver, transmitter cluster are a function of the elevation spreads $\sigma_{\theta_{Rx},c}$ and $\sigma_{\theta_{Tx},c}$,

$$h_{Rx,c} = d_{Rx,c} \tan(\sigma_{\theta_{Rx},c})$$
$$h_{Tx,c} = d_{Tx,c} \tan(\sigma_{\theta_{Tx},c}). \tag{4.37}$$

Similar to the local clusters, the twin cluster is "circular" in the xy-plane. The remaining two axes are set equal to the cluster diameter, (i.e., $a_{Rx,c} = b_{Rx,c} = d_{\tau,c}$). These spans are the same at the transmitter, so that $a_{Tx,c} = b_{Tx,c} = d_{\tau,c}$.

In the above discussion, the cluster position described by the triplet $\mathbf{r}_{X,c,\mathrm{sph}} = (\varphi_{X,c}, \theta_{X,c}, d_{X,c})$ is relative to the position of the receiver or transmitter. To have an absolute reference, $\mathbf{r}_{X,c,\mathrm{sph}}$ must be converted to Cartesian coordinates and added to the location of the receiver or transmitter. Thus, the receiver, transmitter cluster positions $\mathbf{r}_{MS,c}$, $\mathbf{r}_{BS,c}$ are given by

$$\mathbf{r}_{Rx,c} = \mathbf{r}_{Rx} + d_{Rx,c}\left[\cos\theta_{Rx,c} \cdot \cos\varphi_{Rx,c} \quad \cos\theta_{Rx,c} \cdot \sin\varphi_{Rx,c} \quad \cos\theta_{Rx,c}\right]$$
$$\mathbf{r}_{Tx,c} = \mathbf{r}_{Tx} + d_{Tx,c}\left[\cos\theta_{Tx,c} \cdot \cos\varphi_{Tx,c} \quad \cos\theta_{Tx,c} \cdot \sin\varphi_{Tx,c} \quad \cos\theta_{Tx,c}\right]. \tag{4.38}$$

4.5.4.6 *Single-Interaction Clusters*   Single-interaction clusters are placed randomly from the LoS path between the receiver and transmitter, following a Gaussian distribution. They are identical when viewed from the receiver and transmitter. For single-interaction clusters, the cluster distance $d_{Tx,c}$ is defined with respect to the transmitter only and is exponentially distributed, such that

$$d_{Tx,c} = r_{\min} - \sigma_r \log(1 - X_r), \tag{4.39}$$

where $r_{\min}$ is the minimum cluster distance, σ_r is the standard deviation, and $X_r \sim N(0, 1)$.

For single-interaction clusters, the cluster height is the same as seen from the receiver or transmitter (i.e., $h_{Rx,c} = h_{Tx,c}$). The cluster height is a function of $d_{Tx,c}$ and the elevation spread $\sigma_{\theta_{Tx},c}$, such that

$$h_{Tx,c} = d_{Tx,c} \tan(\sigma_{\theta_{Tx},c}). \tag{4.40}$$

Unlike local or twin clusters, a single-interaction cluster has unique spans in the azimuth and delay domains. The azimuth span $b_{Rx,c} = b_{Tx,c}$ and is computed using the azimuth spread $\sigma_{\varphi_{Tx},c}$ as

$$b_{Tx,c} = d_{Tx,c} \tan(\sigma_{\varphi_{Tx},c}). \tag{4.41}$$

The delay span is equal to the cluster diameter

$$a_{Rx,c} = a_{Tx,c} = d_{\tau,c}. \tag{4.42}$$

The cluster azimuth AoD is normally distributed, with the mean being the LoS path between the receiver and transmitter,

$$\varphi_{\text{Tx},c} = \varphi_{\text{Rx,Tx}} + X_{\text{single}} \cdot \sigma_{\varphi_{\text{Tx}},\text{single},c}, \tag{4.43}$$

where $\sigma_{\varphi_{\text{Tx}},\text{single},c}$ is the azimuth spread for single-interaction clusters, and $X_{\text{single}} \sim N(0, 1)$. Note that $\sigma_{\varphi_{\text{Tx}},\text{single},c}$ is different from $\sigma_{\varphi_{\text{Tx}},c}$ for local and twin clusters. The cluster azimuth AoA is identical to the AoD, so that $\varphi_{\text{Rx},c} = \varphi_{\text{Tx},c}$. Note that, in Ref. 91, no distribution is provided for the elevation AoA, AoD for single-interaction clusters in any environment. In the large urban macrocell implementation, $\theta_{\text{Rx},c} = \theta_{\text{Tx},c} = 0$ is assumed.

The absolute cluster position $\mathbf{r}_{\text{Rx},c} = \mathbf{r}_{\text{Tx},c}$ is given by adding the converted relative cluster coordinates to the transmitter position,

$$\mathbf{r}_{\text{Tx},c} = \mathbf{r}_{\text{Tx}} + d_{\text{Tx},c}\big[\cos\theta_{\text{Tx},c} \cdot \cos\varphi_{\text{Tx},c}, \quad \cos\theta_{\text{Tx},c} \cdot \sin\varphi_{\text{Tx},c}, \quad \cos\theta_{\text{Tx},c} \big]. \tag{4.44}$$

4.5.4.7 Multipath Component Placement within Clusters and Multipath Component Power

A fixed number of multipath components $\ell = \{1, \ldots, N_{\text{MPC}}\}$ are normally distributed in each cluster, where N_{MPC} is defined for each environment. The multipath component distribution is the same for all cluster types. We begin by generating a Gaussian random vector $\mathbf{r} = [r_x, r_y, r_z]$, where $r_i \sim N(0, 1)$. Let

$$\mathbf{S}_{X,c} = \begin{bmatrix} a_{X,c} & 0 & 0 \\ 0 & b_{X,c} & 0 \\ 0 & 0 & h_{X,c} \end{bmatrix} \tag{4.45}$$

be the diagonal matrix representing the spans of each cluster for $X \in \{\text{Rx, Tx}\}$. The absolute positions of the ℓth multipath component in cluster c is relative to the cluster position $\mathbf{r}_{X,c}$ and is computed as

$$\mathbf{r}_{X,c,\ell} = \mathbf{r}_{X,c} + \mathbf{r}\mathbf{S}_{X,c}\mathbf{T}(\varphi_{X,c}, \theta_{X,c}), \tag{4.46}$$

where

$$\mathbf{r}_{X,c,\ell} = \begin{bmatrix} x_{X,c,\ell} & y_{X,c,\ell} & z_{X,c,\ell} \end{bmatrix} \tag{4.47}$$

is the multipath component location in Cartesian coordinates. The *rotation matrix* $\mathbf{T}(\varphi, \theta)$ is used to rotate the multipath components toward the receiver or transmitter and is given by

$$\mathbf{T}(\varphi, \theta) = \begin{bmatrix} \cos(\varphi)\cos(\theta) & -\sin(\varphi) & \cos(\varphi)\sin(\theta) \\ \sin(\varphi)\cos(\theta) & \cos(\varphi) & \sin(\varphi)\cos(\theta) \\ -\sin(\theta) & 0 & \cos(\theta) \end{bmatrix}. \tag{4.48}$$

Note that $\mathbf{r}_{X,c,\ell}$ is in Cartesian coordinates. To compute the elements of the H-matrix using the GCM (Section 4.5.1), the multipath component locations need to be converted to polar coordinates. Then, the distances between the multipath components and the receiver, transmitter and the relative AoA, AoD of each multipath component can be computed. Let the difference vector

$$\mathbf{r}_{\Delta X,c,l} = \mathbf{r}_{X,c,l} - \mathbf{r}_X \tag{4.49}$$

for $X \in \{\mathrm{Rx}, \mathrm{Tx}\}$ denote the vector between the multipath component and the receiver, transmitter. The multipath component distances, AoA and AoD are computed as

$$
\begin{bmatrix} \varphi_{X,c,\ell} \\ \theta_{X,c,\ell} \\ d_{X,c,\ell} \end{bmatrix} = \begin{bmatrix} \mathrm{atan}\, 2(\, y_{X,c,\ell}/x_{X,c,\ell}) \\ \mathrm{atan}\, 2\!\left(z_{X,c,\ell} / \sqrt{x_{X,c,\ell}^2 + y_{X,c,\ell}^2}\, \right) \\ \sqrt{x_{X,c,\ell}^2 + y_{X,c,\ell}^2 + z_{X,c,\ell}^2} \end{bmatrix}, \tag{4.50}
$$

where the function $\mathrm{atan}\, 2(\cdot)$ is the four quadrant extension of $\mathrm{atan}(\cdot)$. (For a more detailed discussion of $\mathrm{atan}(\cdot)$, refer to Appendix C.)

The *multipath component delay* $\tau_{c,\ell}$ is a function of the multipath component distance, where, for each multipath component,

$$\tau_{c,\ell} = \frac{d_{\mathrm{Rx},c,\ell} + d_{\mathrm{Tx},c,\ell}}{c_0} + \tau_{\mathrm{link},c}, \tag{4.51}$$

where $\tau_{\mathrm{link},c} = 0$ for local and single-interaction clusters.

4.5.4.8 Multipath Component Power, LoS Attenuation, and Multipath Component Attenuation

The *multipath component power* $P_{c,\ell}$ is Ricean-distributed, with K-factor K_{MPC} such that

$$P_{c,\ell} = |X_{\mathrm{MPC}}\sigma_K + jY_{\mathrm{MPC}}\sigma_K + K_{\mathrm{MPC}}|, \tag{4.52}$$

where σ_K is the standard deviation of the scattered component, and $K_{\mathrm{MPC}} = A_K^2/(2\sigma_K^2)$, where A_K is the amplitude of the LoS component, and X_{MPC}, $Y_{\mathrm{MPC}} \sim N(0, 1)$. Each $P_{c,\ell}$ is normalized with respect to the receiver average power,

$$\overline{P}_{c,\ell} = \frac{P_{c,\ell}}{\sqrt{\frac{1}{N_{\mathrm{MPC}}} \sum_{N_{\mathrm{MPC}}} \sum_{N_c} |P_{c,\ell}|^2}}. \tag{4.53}$$

The *LoS attenuation* A_{LoS} is log-normally distributed with parameters μ_K, σ_K such that

$$A_{\mathrm{LoS}} = \mu_K + 10^{\varepsilon X_{\mathrm{LoS}}\sigma_K}, \tag{4.54}$$

where $\varepsilon = 0.5$.

The *multipath component attenuation* $A_{\text{Att},c,\ell}$ is a function of the fading factor S_c, the cluster power P_c, and the multipath component power $P_{c,\ell}$, where S_c and P_c were computed at (4.25) and (4.24) respectively. For a given c, ℓ,

$$A_{\text{Att},c,l} = \sqrt{P_c \cdot P_{c,\ell} \cdot S_c}. \tag{4.55}$$

The attenuation can also be normalized to its average value,

$$\overline{A}_{\text{Att},c,l} = \frac{A_{\text{Att},c,l}}{\sqrt{\displaystyle\sum_{N_{\text{MPC}}} \sum_{N_c} |A_{\text{Att},c,l}|^2}}. \tag{4.56}$$

4.5.4.9 Computing the Impulse Response

In the final step, the multipath components for all clusters are used to compute the elements of the H-matrix via the discrete form of the GCM presented in Section 4.5.1, extended to include the cluster index c. Let h_{LoS} denote the LoS contribution and $h_{\text{NLoS},c,\ell}$ be the contributions from all multipath components, for $c \in \{1, \dots, N_c\}$, and $\ell \in \{1, \dots, N_{\text{MPC}}\}$, where

$$h_{\text{LoS}}(\tau) = A_{\text{LoS}} e^{j2\pi\tau_0 f_c} \cdot \delta(\tau - \tau_0),$$
$$h_{\text{NLoS},c,\ell}(\tau) = \overline{A}_{\text{Att},c,\ell} e^{j2\pi\tau_{c,l} f_c} \cdot \delta(\tau - \tau_0). \tag{4.57}$$

The LoS contribution to the H-matrix is computed as

$$h_{\text{LoS},m,n}(\tau) = G_m(\varphi_{\text{Rx,Tx}}, \theta_{\text{Rx,Tx}}) h_{\text{LoS}}(\tau) G_n(\varphi_{\text{Rx,Tx}}, \theta_{\text{Rx,Tx}}). \tag{4.58}$$

The NLoS contribution is

$$h_{\text{NLoS},m,n}(\tau) = \sum_{N_c} \sum_{N_{\text{MPC}}} G_m(\varphi_{\text{Rx},c,\ell}, \theta_{\text{Rx},c,\ell}) \cdot h_{\text{NLoS},c,\ell}(\tau)$$
$$\times G_n(\varphi_{\text{Rx},c,\ell}, \theta_{\text{Rx},c,\ell}), \tag{4.59}$$

where $G_m(\varphi, \theta)$, $G_n(\varphi, \theta)$ is the gain of the mth receive, nth transmit antenna, respectively, in the direction of (φ, θ) for $m \in \{1, \dots, M_{\text{Rx}}\}$, and $n \in \{1, \dots, M_{\text{Tx}}\}$. Finally, the H-matrix is computed as the sum of the LoS and NLoS components,

$$h_{m,n} = h_{\text{LoS},m,n} + h_{\text{NLoS},m,n}. \tag{4.60}$$

A summary of the COST 273 model parameters is given in Appendix C.

4.5.5 Features Not Yet Implemented and Omissions

The COST 273 model is broad enough to include more situations than those listed in the previous sections. This section discusses two important features included in the COST 273 framework but left for future implementations. Both of these features deal with the large-scale time-variation in the channel. Specifically, the model includes a mechanism that simulates the appearance and disappearance of clusters from the receiver's view. It also has the ability to simulate long-distance paths, during which some of the cluster parameters become decorrelated.

4.5.5.1 Visibility Regions
In addition to pathloss, when the receiver moves through an environment, the received power varies due to new clusters moving into view, and the occlusion of existing clusters. This process is simulated by surrounding each cluster with a *visibility region*. When the receiver moves within a visibility region, the corresponding cluster is included in the received signal. Each visibility region also includes a *transition region*, in which the power of the associated cluster is transitioned smoothly into and out of view.

Each visibility region is parameterized by the size of the region R_c and the size of the transition region L_c, illustrated in Fig. 4.6. When the receiver enters the visibility region, the cluster power is attenuated by a factor of A_c^2, where

$$A_c = \frac{1}{2} - \frac{1}{\pi} \arctan\left(\frac{2\sqrt{2}d_{\mathrm{Rx},v}}{\lambda L_c}\right). \tag{4.61}$$

The normalized distance $d_{\mathrm{Rx},v}$ is computed as

$$d_{\mathrm{Rx},v} = |\mathbf{r}_{\mathrm{Rx}} - \mathbf{r}_c| + L_c - R_c, \tag{4.62}$$

where $\mathbf{r}_{\mathrm{Rx}}$ is the receiver position, and $\mathbf{r}_c$ denotes the center of the visibility region. Figure 4.7 shows A_c^2 for the case when the receiver moves from the center of the transition region to its edge, and given $R_c = 100$ m, $L_c = 20$ m.

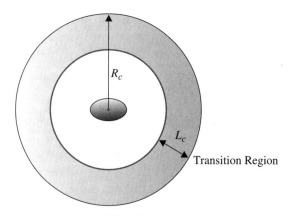

Figure 4.6 COST 273 cluster visibility region.

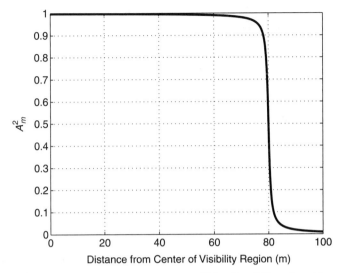

Figure 4.7 A plot of transition region A_c^2 for $R_c = 100\,\text{m}$, $L_c = 20\,\text{m}$.

4.5.5.2 Correlation Distances and Large-Scale Receiver Movement
Consider the case where the receiver is moving through an environment. For the large urban macrocell environment, $\mathbf{r}_{\text{Rx}}$ is varied, and all cluster AoAs, AoDs, and powers are updated accordingly. Recall that the cluster fading factor S_c, as well as the delay, azimuth, and elevation spreads $\sigma_{\tau,c}$, $\sigma_{\varphi_{\text{Rx}},c}$, $\sigma_{\varphi_{\text{Rx}},c}$, $\sigma_{\theta_{\text{Tx}},c}$, $\sigma_{\theta_{\text{Tx}},c}$ are random variables. These parameters decorrelate as the receiver moves relative to its original position. Thus, each of these parameters is associated with a spatial autocorrelation function $R_{X,X'}$, which determines the autocorrelation of each parameter versus distance. The COST 273 model assumes that $R_{X,X'}$ is exponential, that is,

$$R_{X,X'} = e^{-|X-X'|/L_X}, \tag{4.63}$$

where L_X is the autocorrelation distance associated with parameter X, and $X \in \{S, \tau, \varphi_{\text{Rx}}, \varphi_{\text{Tx}}, \theta_{\text{Rx}}, \theta_{\text{Tx}}\}$.

4.5.5.3 Omissions

- Cluster movement: Cluster movement is stochastic, with mean $\mu_{c,v}$ and standard deviation $\sigma_{c,v}$. However, all three environments (macrocell, microcell, and picocell) have $\mu_{c,v} = \sigma_{c,v} = 0$ (91).
- Not all environments implemented: Only the large urban macrocell environment has been implemented (97).
- Not all cluster statistics available: Because of a lack of measurement data, no marginal PDF for the cluster AoA, AoD is specified for the large urban macrocell environment in the COST 273 standard. For example, the

MATLAB implementation assumes that twin clusters are equally probable to arrive/depart from all angles in the azimuth and within 1 rad in elevation, which is a simple assumption. Not all statistics are provided for the single-interaction clusters. For example, values for σ_r and r_{min} are notably absent in all environments.

4.5.6 Disadvantages and Advantages: COST 273

4.5.6.1 Disadvantages

- Model complexity: Here, the problem of complexity does not refer to number of parameters but rather to the fact that the model itself has many working parts, and many of these change from one environment to the next. Keeping track of all the parameters, and implementing a program that is general enough to include all environments, is quite a task.

- Stochastic parameter complexity: Some of the stochastic parameters are not easy to compute from measured data or are very sensitive to calibration error. For example, angular power measurements require a great deal of care when calibrating for power and antenna gain in the azimuth and elevation. This is hard to do in practice and may change under certain environmental conditions. However, the GCM, on which the model is based, is general enough so that certain parameters can be omitted to simplify the model. For example, the elevation parameters can be omitted from the model to get an azimuth-only equivalent channel. This relaxes the requirement for elevation-dependent parameters such as antenna gain, cluster AoA, AoD, and so forth.

- Model specificity: The model does not accurately represent any single channel but rather represents the average behavior of many similar environments, such as the large urban macrocell. This is good for device designers that have to make their product work in a multitude of environments. However, this can be bad if one desires accurately to reproduce a single channel, such as a specific path between two buildings on a campus.

- Not all aspects implemented: As discussed above, not all aspects of the COST 273 model have been implemented. In addition, the parameter list is incomplete for all environments.

4.5.6.2 Advantages

- Physical interpretation and wide applicability: As with most standardized models, the parameters, and the model itself, both hold a relatively straightforward physical interpretation. This means that, even without hard measurement data, certain parameters can be changed to reflect different environments and the effects investigated. In addition, when the parameters are computed from measured data, they often have meaning that can be tied directly to the measurement environment, such as the density of scatterers around each link-end. The model is general enough to apply to many environments. A new environment can be specified by using a sounder to measure the impulse response of a number of exemplar channels. The parameters for the new environment can be computed from the data.

4.6 THE RANDOM CLUSTER MODEL

In Ref. 80, Czink presents a cluster model based on a refinement of the COST 273 model. There are two features that distinguish the RCM from the COST 273 model: (1) cluster parameters are identified *automatically* from measured data, and (2) the clusters are completely characterized using a *multivariate PDF* called the *environment PDF*. Unlike the COST 273 channel model, the RCM focuses on automating the process of identifying cluster parameters from measured data and streamlining the simulation procedure.

Similar to COST 273, The RCM is a system-level cluster model. It is suitable for use in evaluating algorithmic and transceiver design trade-offs in specific environments. Unlike the COST 273 model, however, the RCM characterizes each environment using the environment PDF. The model does not change from one environment to the next; only the environment PDF changes. This reduces the implementation complexity considerably. The environment PDF is computed automatically from measured data. This means that the resulting model is coupled very closely to the measured environment, and even to the specific path of the measurement device through the environment. This makes it difficult to include the RCM in a standard; the specificity of the environment PDF makes it hard to describe the average behavior of many channels in a similar environment.

Instead of visibility regions, the RCM introduces a smoothly time-varying *cluster birth–death process* to simulate movement, where cluster powers are gradually increased, exist for a period, and gradually fade out at some later point. The parameters governing cluster movement are computed automatically from measured data using a novel *joint clustering and tracking framework*. We discuss both of these concepts further in the following section.

In Chapter 6, we present a performance analysis of the RCM using real-life data.

4.6.1 General Description

Figure 4.8 presents an overview of the process involved in generating exemplar channels using the RCM.

The RCM consists of three major components: external parameters, a parametric channel model, and a system model. The environment PDF Θ_{env} is a stochastic description of the clusters, and it is estimated from measured data. Using Θ_{env}, the parametric channel model is used to generate a set of *cluster* and *multipath component parameter sets* Θ_c and $\Theta_{c,\ell}$ respectively. The system model incorporates system parameters Θ_{sys}, such as antenna pattern, number of antennas, and so forth, and Θ_c and $\Theta_{c,\ell}$, to produce a time variant, wideband H-matrix $\mathbf{H}(t, f)$.

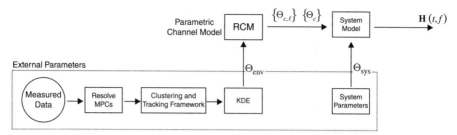

Figure 4.8 The RCM general framework, illustrating the steps from measured data to time-variant channel matrix. KDE, kernal density estimator.

4.6.1.1 Cluster Parameters

The parametric channel model describes the time-variant wideband H-matrix as

$$\mathbf{H}(t, f, \mathbf{\Theta}_{\text{env}}) = \sum_{c=1}^{N_c} \mathbf{H}_c(t, f, \mathbf{\Theta}_c), \tag{4.64}$$

where $\mathbf{H}_c(t, f, \mathbf{\Theta}_c)$ is the contribution from the cth cluster. In total, there are N_c clusters.

The set $\{\mathbf{\Theta}_c\}$ uniquely describes all N_c clusters in a given snapshot, where the *snapshot period* is denoted Δt_s. Each $\mathbf{\Theta}_c$ consists of a number of parameters. These parameters are used to identify the cluster location in parameter space, spread, power, their evolution in time, and the number of multipath components associated with each cluster. As with the COST 273 model, the location of the cluster is defined in the double-directional azimuth-elevation-delay space. Accordingly, the location parameters of each cluster are the *mean azimuth angles* $\overline{\varphi}_{\text{Rx},c}$, $\overline{\varphi}_{\text{Tx},c}$, *mean elevation angles* $\overline{\theta}_{\text{Rx},c}$, $\overline{\theta}_{\text{Tx},c}$, and *mean delay* $\overline{\tau}_c$. Each of these parameters is associated with a receiver *spread*, $\sigma_{\varphi_{\text{Rx}},c}$, $\sigma_{\varphi_{\text{Tx}},c}$, $\sigma_{\theta_{\text{Rx}},c}$, $\sigma_{\theta_{\text{Tx}},c}$, and $\sigma_{\tau,c}$, respectively. The *cluster power* is defined by $\sigma_{\gamma,c}^2$, and the total *snapshot power* by ρ_c, where we refer to the snapshot in which cluster c exists.

Time-variation is simulated by linearly varying the cluster position and power parameters from one snapshot to the next. Accordingly, the cluster *rate-of-change* parameters are $\Delta\overline{\varphi}_{\text{Rx},c}$, $\Delta\overline{\varphi}_{\text{Tx},c}$, $\Delta\overline{\theta}_{\text{Rx},c}$, $\Delta\overline{\theta}_{\text{Tx},c}$, $\Delta\overline{\tau}_c$, and $\Delta\sigma_{\gamma,c}^2$. Along with cluster movement, the cluster power is set to fade in, "live" for a predetermined period, and fade out gracefully. This process is referred to as the cluster birth–death process and is discussed below. The *cluster lifetime* is given by Λ_c, which is defined as a multiple of the *cluster lifetime interval* Δt_Λ, which in turn is a multiple of Δt_s. The fade-in and fade-out processes are governed by the *cluster attenuation parameter* $|\gamma_{\text{att}}|^2$.

4.6.1.2 Multipath Component Parameters

Each cluster is composed of $N_{c,\ell}$ multipath components. For a given c, the set $\{\mathbf{\Theta}_{c,\ell}\}$ for all ℓ uniquely identifies the location and power of all multipath components within that cluster. The contribution of the ℓth multipath component is defined as

$$\mathbf{H}_c(t, f, \mathbf{\Theta}_c) = \sum_{\ell=1}^{N_{c,\ell}} \mathbf{H}_{c,\ell}(t, f, \mathbf{\Theta}_{c,\ell}). \tag{4.65}$$

The multipath component parameter set $\{\Theta_{c,\ell}\}$ consists of the *multipath component azimuth angles* $\varphi_{\mathrm{Rx},c,\ell}$ and $\varphi_{\mathrm{Tx},c,\ell}$, *multipath component elevation angles* $\theta_{\mathrm{Rx},c,\ell}$, $\theta_{\mathrm{Tx},c,\ell}$, the *multipath component delay* $\tau_{c,\ell}$, and the *complex multipath component gain* $\gamma_{c,\ell}$.

4.6.1.3 The Environment PDF

One of the unique features of the RCM is that it completely describes the behavior of the channel via a single multivariate random variable. The environment PDF Θ_{env} is the joint distribution of all cluster parameters, Θ_c, that is,

$$\Theta_{\mathrm{env}} = p(\Theta_c)$$

$$= p(\overline{\varphi}_{\mathrm{Rx}}, \overline{\varphi}_{\mathrm{Tx}}, \overline{\theta}_{\mathrm{Rx}}, \overline{\theta}_{\mathrm{Tx}}, \overline{\tau}_c, \sigma_{\varphi_{\mathrm{Rx}},c}, \sigma_{\varphi_{\mathrm{Tx}},c}, \sigma_{\theta_{\mathrm{Rx}},c}, \sigma_{\theta_{\mathrm{Tx}},c}, \sigma_{\tau,c}, \sigma_{\gamma,c}^2, \rho_c,$$

$$N_c, N_{c,\ell}, \Delta\overline{\varphi}_{\mathrm{Rx},c}, \Delta\overline{\varphi}_{\mathrm{Tx},c}, \Delta\overline{\theta}_{\mathrm{Rx},c}, \Delta\overline{\theta}_{\mathrm{Tx},c}, \Delta\overline{\tau}_c, \Delta\sigma_{\gamma,c}^2, \Lambda_c). \qquad (4.66)$$

Θ_{env} is computed directly from measured data, using, for example, a *kernel density estimator* (100). In general, a kernel density estimator is used to estimate the PDF of a random variable given a sample data set. In Ref. 100, Ihler presents a kernel density estimator toolbox for MATLAB that can be used to estimate multivariate PDFs. Because Θ_{env} is generated from a data set using a kernel density estimator, it is representative of that data set only and may not be an accurate characterization of other environments. This multivariate PDF description is useful in that it closely reflects the true environment; in practice, many cluster parameters are correlated. For example, as the number of clusters increases, we would expect their spreads to decrease. It also provides a relatively easy way of generating more exemplar channels.

We can use Θ_{env} to generate an ensemble of exemplar channels consisting of clusters whose behavior mimics that of the measured channel. We discuss this process further in the next section.

4.6.1.4 Generating Exemplar Channels Using the RCM

Once Θ_{env} has been specified, the RCM can be used to generate a set of time-varying clusters. The following procedure is repeated for every snapshot.

1. Determine the number of clusters in the current snapshot N_c by marginalizing Θ_{env} and choosing N_c from the marginal PDF $p(N_c)$. We marginalize Θ_{env} by integrating all variables except N_c out of the multivariate PDF Θ_{env}.
2. Determine the snapshot power ρ_c by marginalizing Θ_{env}, conditioned on the N_c generated in the previous step, and choosing from $p(\rho_c|N_c)$.
3. Condition Θ_{env} on N_c, ρ_c to get the conditional PDF $p(\Theta_c|N_c, \rho_c)$. Using this, generate a new set of cluster parameters $\{\Theta_c\}$ for $c = \{1, \ldots, N_c\}$.
4. For each Θ_c generated in the previous step, generate a set of $N_{c,\ell}$ path parameter sets $\Theta_{c,\ell}$. The path-location parameters are generated according to a Gaussian

distribution in accordance with

$$
\begin{aligned}
\varphi_{\text{Rx},c,\ell} &\sim N(\overline{\varphi}_{\text{Rx}}, \sigma^2_{\varphi_{\text{Rx}},c}) \\
\varphi_{\text{Tx},c,\ell} &\sim N(\overline{\varphi}_{\text{Tx}}, \sigma^2_{\varphi_{\text{Tx}},c}) \\
\theta_{\text{Rx},c,\ell} &\sim N(\overline{\theta}_{\text{Rx}}, \sigma^2_{\theta_{\text{Rx}},c}) \\
\theta_{\text{Tx},c,\ell} &\sim N(\overline{\theta}_{\text{Tx}}, \sigma^2_{\theta_{\text{Tx}},c})
\end{aligned}
\tag{4.67}
$$

where $N(\cdot,\cdot)$ is the Gaussian distribution, the first element of the argument stands for the mean, and the second element stands for the variance. In order for these to be meaningful, the angular values are mapped to the interval $(-\pi, \pi]$. The complex path-gain $\gamma_{c,\ell} = |\gamma_{c,\ell}| \angle \gamma_{c,\ell}$ is computed according to

$$
|\gamma_{c,\ell}| = \sqrt{\frac{\sigma^2_{\gamma,c}}{|\gamma_{\text{att}}|^2 N_{c,\ell}}},
\tag{4.68}
$$

$$
\angle \gamma_{c,\ell} = U(-\pi, \pi)
$$

where $U(\cdot,\cdot)$ is the uniform distribution.

5. Using the system model (introduced in the following section), compute a new $\mathbf{H}(t, f)$.

4.6.1.5 The System Model

The system model uses a subset of parameters from Θ_c and $\Theta_{c,\ell}$, and a set of external parameters that describe the measurement system, to generate wideband H-matrix $\mathbf{H}(t, f)$. The external parameters are the $M_{\text{Rx}} \times 1$ receive and $M_{\text{Tx}} \times 1$ transmit *array steering vectors* $\mathbf{a}_{\text{Rx}}(\varphi_{\text{Rx}}, \theta_{\text{Rx}})$, $\mathbf{a}_{\text{Tx}}(\varphi_{\text{Tx}}, \theta_{\text{Tx}})$, respectively. Note that the array steering vectors implicitly include the number of antennas, the position of the antenna elements in each array, and the complete azimuth-elevation pattern of each antenna. Given the steering vectors, $\mathbf{H}(t, f)$ is computed for each snapshot as

$$
\mathbf{H}(t, f) = \sum_{c=1}^{N_c} \sum_{\ell=1}^{N_{c,\ell}} \gamma_{c,\ell} \mathbf{a}_{\text{Rx}}(\varphi_{\text{Rx}}, \theta_{\text{Rx}}) \mathbf{a}_{\text{Tx}}^T(\varphi_{\text{Tx}}, \theta_{\text{Tx}}) e^{-j2\pi f \tau_{c,\ell}}.
\tag{4.69}
$$

Note that the frequency range and resolution are dependent on two external parameters, the bandwidth B and the maximum resolvable delay $\tau_{\max}$, of the measurement system. The frequency resolution is determined by $\Delta f = 1/\tau_{\max}$ and the number of frequency bins $N_{\Delta f} = B/\Delta f$. Time in the system model is implied in the cluster parameters; the cluster parameters are updated for each snapshot according to the cluster birth–death process, as described in the following section.

4.6.1.6 Smoothly Time-Varying Channels and the Cluster Birth–Death Process

In the case where the RCM is used to generate a smoothly time-variant channel, the path parameters are updated for every snapshot for the lifetime of the related cluster. The RCM only considers the case where the *transmitter* moves through the environment. For the next snapshot, the locations of all paths are updated in parameter space according to

$$
\begin{aligned}
\varphi_{\text{Rx},c,\ell}^{(n+1)} &= \varphi_{\text{Rx},c,\ell}^{(n)} + \Delta\overline{\varphi}_{\text{Rx},c}d_\lambda \\
\varphi_{\text{Tx},c,\ell}^{(n+1)} &= \varphi_{\text{Tx},c,\ell}^{(n)} + \Delta\overline{\varphi}_{\text{Tx},c}d_\lambda \\
\theta_{\text{Rx},c,\ell}^{(n+1)} &= \theta_{\text{Rx},c,\ell}^{(n)} + \Delta\overline{\theta}_{\text{Rx},c}d_\lambda \\
\theta_{\text{Tx},c,\ell}^{(n+1)} &= \theta_{\text{Tx},c,\ell}^{(n)} + \Delta\overline{\theta}_{\text{Tx},c}d_\lambda \\
\tau_{c,\ell}^{(n+1)} &= \tau_{c,\ell}^{n} + \Delta\overline{\tau}_{c}d_\lambda
\end{aligned}
\tag{4.70}
$$

where $d_\lambda = v_{\text{Tx}}\Delta t_s$ is the distance the transmitter has traveled since the last snapshot, measured in terms of wavelengths, v_{Tx} is the transmitter speed in wavelengths per second, and superscripts (n) and $(n+1)$ denote the current and updated parameter, respectively. The magnitude of the path gains $|\gamma_{c,\ell}|$ evolve according to

$$
|\gamma_{c,\ell}^{(n+1)}|^2(dB) = |\gamma_{c,\ell}^{(n)}|^2(dB) + \Delta\sigma_{\gamma,c}^2 d_\lambda,
\tag{4.71}
$$

while the phase $\angle\gamma_{c,\ell}$ remains unchanged.

In addition to the cluster motion over time, there are several parameters that determine when a cluster is born, how long it lives, and when it dies. In addition, clusters do not simply appear and disappear; the cluster birth-process involves gradually increasing the cluster power over a fixed duration, whereas cluster death-process involves decreasing the cluster power gradually over the same duration. Cluster births and deaths are reevaluated every cluster lifetime interval, Δt_Λ. Cluster births are governed by $p(x_{\text{birth}})$, which describes the cluster birth PDF, and the cluster lifetime Λ_c. On each Δt_Λ, we draw a new x_{birth} from $p(x_{\text{birth}})$, and we simultaneously remove clusters at the end of their lifetimes. The total N_c is updated accordingly. The RCM uses a simple linear-in-dB method of fading clusters in and out; clusters are faded in by decreasing $|\gamma_{\text{att}}|^2$ from -40 to 0 dB every Δt_s for one Δt_Λ. Similarly, clusters are faded out by increasing $|\gamma_{\text{att}}|^2$ from 0 to -40 dB over the same period.

4.6.2 Determining the Environment PDF

The task of generating Θ_{env} from measured data is far from trivial. Arguably, the largest effort in modeling real-life channels is spent in computing Θ_{env} from the measured data. This process can roughly be divided into four steps:

1. A super-resolution algorithm, such as SAGE (95) or ISIS (96), is used to identify all multipath components in the data. (This is a common first step for most modern clustering algorithms, including COST 273. Often, it is the most time consuming one.)

2. The paths are grouped into clusters using a clustering algorithm, such as KPowerMeans.

3. A *Kalman filter* (101, 102) is used to track the cluster centroid positions and speeds over time.

4. Finally, statistical methods such as the kernel density estimator (see Ref. 100 for a MATLAB code example) are used to determine the multivariate Θ_{env} from the tracked parameters.

In the following, we briefly comment on each step (80).

4.6.2.1 Joint Clustering and Tracking Framework

The process of determining Θ_{env} first involves the use of a "joint clustering-and-tracking framework" to extract cluster parameters from time-variant channel data. This framework is iterative. First, a Kalman filter (101, 102) is used to track the cluster centroid positions and speeds over time. The Kalman filter also predicts the cluster positions in the next snapshot. An *initial-guess* routine takes the Kalman-predicted cluster centroids, and a new snapshot, to produce an initial guess of the cluster centroids for the clustering algorithm. The KPowerMeans clustering algorithm is used to produce the final estimate for the cluster parameters. These new cluster parameters are used to update the Kalman filter, and the process is repeated for each new snapshot.

4.6.2.2 KPowerMeans Clustering Algorithm

As previously mentioned, the task of clustering is, in general, complex. Visual clustering is difficult because, in many cases, clusters do not to have well-defined shapes, and clusters may overlap. It is also difficult in general to identify clusters in the delay-domain. The KPowerMeans clustering algorithm is an automatic method of clustering multipath components in a given snapshot. It does not track clusters over time, however. One of the important contributions of Ref. 80 was to use a Kalman filter to track the clusters over time and maintain continuity between snapshots.

The KPowerMeans clustering algorithm initially needs to know N_c. In the case where this is not given, the KPowerMeans algorithm works with a number of randomly picked initial starting points and converges to a solution. The solution with the smallest *distance metric*, discussed below, is chosen.

The value N_c is also inferred directly from measured data using an initial-guess procedure involving a power-based cutoff criterion (80). While the initial guess routine tries to find as many distinct clusters as possible, the KPowerMeans algorithm produces compact clusters.

The KPowerMeans algorithm uses a distance metric based on the *multipath component distance* (103), weighted by the path gain. The KPowerMeans algorithm is iterative. Because weights are assigned to each multipath component distance based on the path gain, the KPowerMeans algorithm clusters multipath component toward centroids with greater power. The algorithm minimizes distance from cluster centers, which has the effect of minimizing cluster angular and delay spreads. Once the clusters have been found in a particular data set, the *power threshold criterion* is used to verify the accuracy of the estimate; a cluster can only exist when its power is greater than a given ratio of the total snapshot power. If this condition is not satisfied, the clustering algorithm must be repeated with a lower initial N_c. Using synthetic data generated from the COST 259 SCM, the KPowerMeans algorithm is shown to be more accurate than previous clustering algorithms (80).

4.6.2.3 Estimating the Probability Density Function of Cluster Parameters
As previously mentioned, a kernel density estimator (see Ref. 100 for a MATLAB code example) is used to estimate the environment PDF. The environmental PDF is written as the sum of kernels $K(\boldsymbol{\theta}_c, \boldsymbol{\mu}_{\Theta_i}, \mathbf{C}_{\Theta_i})$ (104),

$$\boldsymbol{\Theta}_{\text{env}} = f(\boldsymbol{\theta}_c) = \frac{1}{N_c} \sum_{i=1}^{N_c} K(\boldsymbol{\theta}_c, \boldsymbol{\mu}_{\Theta_i}, \mathbf{C}_{\Theta_i}), \tag{4.72}$$

where N_c is the number of clusters, and $\boldsymbol{\mu}_{\Theta_i}$ and $\mathbf{C}_{\Theta_i}$ are the mean and covariance of the ith kernel, respectively. Each $K(\boldsymbol{\theta}_c, \boldsymbol{\mu}_{\Theta_i}, \mathbf{C}_{\Theta_i})$ is modeled as a Gaussian PDF,

$$K(\boldsymbol{\theta}_c, \boldsymbol{\mu}_{\Theta_i}, \mathbf{C}_{\Theta_i}) = \frac{1}{(2\pi)^{D/2}|\mathbf{C}_{\Theta_i}|^{1/2}} e^{-\frac{1}{2}(\boldsymbol{\theta}_c - \boldsymbol{\mu}_{\Theta_i})^T \mathbf{C}_{\Theta_i}^{-1}(\boldsymbol{\theta}_c - \boldsymbol{\mu}_{\Theta_i})}, \tag{4.73}$$

where $D = 21$ denotes the number of cluster parameters.

4.6.3 Advantages and Disadvantages: The RCM

4.6.3.1 Advantages

- Time-variation: The RCM focuses on tracking clusters over time, thus accurately simulating channel time-variation.
- Double-directional: As with the COST 273 model, the RCM is based on the double-directional description of the channel. The double-directional cluster model is an intuitive and physically meaningful way of describing path-based energy propagation.
- System independence: Like the COST 273 model, the RCM is system-independent and models the effects of the multipath component alone. This means that trade-offs using different transceiver architectures, including different antennas, can be investigated while keeping the channel fixed.
- Ease of use: Once the environmental PDF $\boldsymbol{\Theta}_{\text{env}}$ has been computed, producing exemplar channels is relatively simple; the channel is completely characterized by a multivariate PDF.

4.6.3.2 Disadvantages

- Parameter computation complexity: Computing $\boldsymbol{\Theta}_{\text{env}}$ is not a trivial task. It involves several relatively complex stages to obtain the final PDF.

- "God" parameters: In addition, the definition of a "cluster" is still not cast in stone; there are several parameters in the clustering and tracking framework that are heuristically optimized and can, for example, affect how the paths are clustered together. Different parameters will yield different results.

- Model specificity: The path parameters are specific to the measured path, and the simulated channel will be specific to that particular path. Thus, given one Θ_{env}, it would be difficult to vary the parameters to produce simulation results for a given environment.

- Indeterminate number of parameters: The total number of parameters for the model is not fixed and changes with every snapshot. The total number of parameters is especially dependent on the number and "size" of clusters in the channel; more clusters means more paths, and bigger clusters require more paths to be properly represented. Each of these parameters is heuristically optimized. Because the cluster parameters are random variables drawn from Θ_{env}, the complexity of the model is also random and presumably depends on the multipath richness of the measured channel.

4.7   SUMMARY AND DISCUSSION

In this chapter, we reviewed some of the fundamentals behind cluster models. Clusters are defined to be a group of multipath components with similar properties, such as AoA, delay, and so forth. Although the concept of a cluster is simple, the exact definition varies in the literature. Until recently, clustering multipath components was done visually, which led to the most arbitrary definition of a cluster.

Clusters were first identified visually in the delay-domain. The Saleh–Valenzuela model describes the arrival times and PDP for clusters in a wideband channel. The model includes only a few measured parameters and is relatively simple. However, the model holds for a great many channels and is still widely used today. The model has since been extended to include the angular domains, both at the transmitter and the receiver.

The COST 273 model is one of the most comprehensive cluster models in existence today. It was the culmination of a large European cooperative effort between many academic and industrial institutions. The model is general enough in that it has the ability to model many different environments accurately. However, this comes at the price of increased complexity. There are numerous stochastic parameters that are dependent on the environment. Overall, the model implementation is not trivial. Simply put, "there is no free lunch"; for every gain we make, there is a price to pay.

The RCM improves on the COST 273 model in several key areas. First, the model is much easier to use. By using a multivariate PDF to characterize the environment, instead of a set of external and stochastic parameters, the RCM streamlines the process of generating exemplar channels. The RCM also automates the process of identifying the time-variant cluster parameters from the measured data. The novel clustering and tracking framework combines cluster identification and cluster tracking. This removes much of the ambiguity in clustering and allows for the accurate measurement of time-variant parameters from the data. Compared with the COST 273 model, the RCM complexity is mostly in identifying cluster parameters from the data. The clustering and tracking framework is not trivial and involves several heuristically optimized parameters.

4.8 NOTES AND REFERENCES

The practice of publishing MATLAB implementations of channel models (and other research-related material) on the Web has become rather widespread. For those who know MATLAB and wish to learn more about some of the models discussed in this chapter, among other popular cluster models, the following lists some valuable references. The scripts are especially useful to those wishing to generate synthetic channel data to test algorithms, transceiver designs, or even other channel models.

- A MATLAB implementation of the COST 273 model (97).
- The IEEE 802.15.3a ultra wide band model and final report, which includes an extended version of the Saleh–Valenzuela model (92).
- A MATLAB implementation of the RCM. http://userver.ftw.at/~nczink
- The COST 207 model, used in the development of GSM, has already been implemented in the MATLAB Communications Toolbox, version 4.2.
- A MATLAB implementation of the 3GPP SCM model is available at the MATLAB Web site (105).
- The MATLAB implementation of the WINNER II model can be downloaded (79).

5

CHANNEL SOUNDING

5.1 INTRODUCTION

The ultimate goal of any channel model is to simulate the behavior of real-life channels. At some point, this implies that the channel itself must be measured and the channel model performance compared using real-life data. Channel modeling is thus closely coupled with *channel sounding*. We measure the channel using a *channel sounder* and a variety of techniques. The technique depends on the type of channel we wish to measure.

This chapter covers some theoretical and practical aspects of channel sounding. We begin by describing an example of a 4×4 MIMO channel sounder, McMaster's *wideband MIMO software defined radio* (WMSDR). The WMSDR was designed to be used to gather real-life channel estimation data. The data can, in turn, be used to develop communication algorithms or channel models.

Here, we discuss the theory behind various channel sounding techniques, leading from *periodic pulse sounding*, probably the simplest form of channel sounding, to wideband MIMO channel sounding techniques. Two channel sounding techniques are of particular importance, as they are popular in the literature today; these are the *digital matched filtering* and *sampled spectrum* techniques. Digital matched filtering is a technique that can be used to obtain calibrated estimates of the wideband MIMO channel. We discuss correlative channel sounding and receiver calibration. We then

Multiple-Input, Multiple-Output Channel Models: Theory and Practice. By Nelson Costa and Simon Haykin
Copyright © 2010 John Wiley & Sons, Inc.

move on to discuss sampled spectrum channel sounding and *switched-array* architectures, which significantly reduce the complexity of a wideband MIMO sounder.

Unlike switched-array sounders, the WMSDR is capable of transmitting digital
data across all four transmitters simultaneously. Transmitting digital data, however,
introduces more complexity; the receiver has to be *synchronized* to the receiver to
recover the transmitted data. In the second half of the chapter, we discuss several
important practical considerations in the transmission and reception of digital signals;
namely, timing, carrier, and phase recovery.

5.2 THE WMSDR

To introduce terminology and concepts used throughout the rest of the chapter, we present a simple example that illustrates the WMSDR theory of operation. The example
serves to illustrate how one can encode, transmit, receive, and process a digital signal
using a pseudo-noise sequence (PN sequence) sounder. The steps in the example are
closely linked to discussions in the rest of the chapter. The steps described here were
used during all WMSDR measurement campaigns, including the one described in
Chapter 6. Elements of this process are also used in all digital correlative sounders,
as well as in digital radios in general.

5.2.1 Transmission

1. Generate Data

Consider the case where we wish to transmit the four sequences:

$$
\begin{aligned}
\text{Transmitter 1: } & m_1[i] = \{1, 2, 3, 4\} \\
\text{Transmitter 2: } & m_2[i] = \{4, 1, 2, 3\} \\
\text{Transmitter 3: } & m_3[i] = \{3, 4, 1, 2\} \\
\text{Transmitter 4: } & m_4[i] = \{2, 3, 4, 1\}.
\end{aligned}
\tag{5.1}
$$

Each sequence consists of $L = 4$ *symbols*, with each symbol being chosen from
an alphabet of size $S = 4$. The WMSDR transmitter is equipped with four independent *transmit chains*, which means it can transmit all four sequences simultaneously. Here, a *chain* refers to the signal path from digital baseband to radio
frequency (RF).

2. Encode Using Signal Constellation

Next, a signal constellation is chosen, and each symbol is encoded as a constellation
point. For the sequence above, at least $b_n = \log_2 4 = 2$ bits/symbol are required. In
this case, the message can be encoded most easily using a *quadrature phase-shift
keyed* (QPSK) constellation, illustrated in Fig. 5.1. Thus, the four message symbols

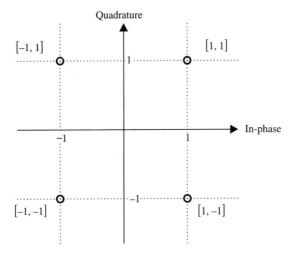

Figure 5.1 QPSK constellation.

can be written as

$$
\begin{aligned}
s_1[i] &= \{[1, 1], [-1, 1], [-1, -1], [1, -1]\} \\
s_2[i] &= \{[1, -1], [1, 1], [-1, 1], [-1, -1]\} \\
s_3[i] &= \{[-1, -1], [1, -1], [1, 1], [-1, 1]\} \\
s_4[i] &= \{[-1, 1], [-1, -1], [1, -1], [1, 1]\},
\end{aligned}
\tag{5.2}
$$

where $s_n[i]$ is the sequence at the nth transmit antenna. We refer to each $s_n[i]$ as a *complex chip* as each consists of an in-phase and quadrature component.

3. Encode Constellation as Interleaved In-phase and Quadrature Binary Numbers

A *quadrature modulator* consist of an *in-phase* and *quadrature* branch (106). By assigning values to each branch, we can specify a point on the *in-phase–quadrature plane* (IQ-plane). Each chip in the transmit sequences consists of a real and an imaginary component. The real and imaginary components are assigned to the in-phase and quadrature branches, respectively. The result is that we separate $s_n[i]$ into its *in-phase* and *quadrature chips*, $s_{n,I}[i]$ and $s_{n,Q}[i]$, respectively, for $n \in \{1, \ldots, 4\}$,

$$
\begin{aligned}
s_{1,I}[i] &= \{+1, -1, -1, +1\}, s_{1,Q}[i] = \{+1, +1, -1, -1\} \\
s_{2,I}[i] &= \{+1, +1, -1, -1\}, s_{2,Q}[i] = \{-1, +1, +1, -1\} \\
s_{3,I}[i] &= \{-1, +1, +1, -1\}, s_{3,Q}[i] = \{-1, -1, +1, +1\} \\
s_{4,I}[i] &= \{-1, -1, +1, +1\}, s_{4,Q}[i] = \{+1, -1, -1, +1\},
\end{aligned}
\tag{5.3}
$$

where the subscripts I and Q denote the in-phase and quadrature chips, respectively.

In order to facilitate digital-to-analog conversion, each chip is converted to a binary number. The WMSDR transmitter is capable of representing each chip as an 8-bit binary number ranging from 0 to 255, with 0 representing the largest negative value and 255 the largest positive value. Thus, we encode each chip using the map $-1 = [00000000]$, $+1 = [11111111]$. The sequences are represented in binary as

$$\begin{aligned}
s_{1,\mathrm{I}}[i] &= \{11111111,\, 00000000,\, 00000000,\, 11111111\} \\
s_{1,\mathrm{Q}}[i] &= \{11111111,\, 11111111,\, 00000000,\, 00000000\} \\
s_{2,\mathrm{I}}[i] &= \{11111111,\, 11111111,\, 00000000,\, 00000000\} \\
s_{2,\mathrm{Q}}[i] &= \{00000000,\, 11111111,\, 11111111,\, 00000000\} \\
s_{3,\mathrm{I}}[i] &= \{00000000,\, 11111111,\, 11111111,\, 00000000\} \\
s_{3,\mathrm{Q}}[i] &= \{00000000,\, 00000000,\, 11111111,\, 11111111\} \\
s_{4,\mathrm{I}}[i] &= \{00000000,\, 00000000,\, 11111111,\, 11111111\} \\
s_{4,\mathrm{Q}}[i] &= \{11111111,\, 00000000,\, 00000000,\, 11111111\}.
\end{aligned} \tag{5.4}$$

The $s_{n,\mathrm{I}}[i]$ and $s_{n,\mathrm{Q}}[i]$ are then interleaved in time to reduce the number of bits transmitted in parallel per clock cycle. The resulting signal, $s_{n,\mathrm{int}}[i]$, is shown in Fig. 5.2.

4. Convert to In-phase and Quadrature Analog Signals

The next step is to convert each chip into an in-phase and quadrature *transmit pulse*. This is done via a quadrature *digital-to-analog converter* (DAC). The WMSDR transmitter is equipped with four quadrature DACs total, one for each transmit chain. The baseband portion of the transmit chain is illustrated in Fig. 5.3.

In each quadrature DAC, the interleaved sequences $s_{n,\mathrm{int}}[i]$ are first de-interleaved into $s_{n,\mathrm{I}}[i]$ and $s_{n,\mathrm{Q}}[i]$. The in-phase and quadrature sequences are then converted to separate analog signals $\hat{s}_{n,\mathrm{I}}(t)$ and $\hat{s}_{n,\mathrm{Q}}(t)$, respectively. The length of each pulse is determined by a sample clock, whose period is T_{Tx}. Note that at the output of the quadrature DAC, the transmitted sequence is represented as a series of square-shaped pulses. This is a very bandwidth-inefficient representation. Thus, we use passive pulse shaping filters at the output of the quadrature DAC to reduce the bandwidth of the sequence at baseband. The results $s_{n,\mathrm{I}}(t)$ and $s_{n,\mathrm{Q}}(t)$ represent the

$$s_{1,\mathrm{int}}[i]\left\{\begin{array}{ll}
11111111 & s_{1,\mathrm{I}}[1] \\
11111111 & s_{1,\mathrm{Q}}[1] \\
00000000 & s_{1,\mathrm{I}}[2] \\
11111111 & s_{1,\mathrm{Q}}[2] \\
00000000 & s_{1,\mathrm{I}}[3] \\
00000000 & s_{1,\mathrm{Q}}[3] \\
11111111 & s_{1,\mathrm{I}}[4] \\
00000000 & s_{1,\mathrm{Q}}[4]
\end{array}\right\}
\quad
s_{2,\mathrm{int}}[i]\left\{\begin{array}{ll}
00000000 & s_{2,\mathrm{I}}[1] \\
11111111 & s_{2,\mathrm{Q}}[1] \\
11111111 & s_{2,\mathrm{I}}[2] \\
11111111 & s_{2,\mathrm{Q}}[2] \\
11111111 & s_{2,\mathrm{I}}[3] \\
00000000 & s_{2,\mathrm{Q}}[3] \\
00000000 & s_{2,\mathrm{I}}[4] \\
00000000 & s_{2,\mathrm{Q}}[4]
\end{array}\right\}
\quad
s_{3,\mathrm{int}}[i]\left\{\begin{array}{ll}
00000000 & s_{3,\mathrm{I}}[1] \\
11111111 & s_{3,\mathrm{Q}}[1] \\
00000000 & s_{3,\mathrm{I}}[2] \\
11111111 & s_{3,\mathrm{Q}}[2] \\
11111111 & s_{3,\mathrm{I}}[3] \\
00000000 & s_{3,\mathrm{Q}}[3] \\
11111111 & s_{3,\mathrm{I}}[4] \\
00000000 & s_{3,\mathrm{Q}}[4]
\end{array}\right\}
\quad
s_{4,\mathrm{int}}[i]\left\{\begin{array}{ll}
00000000 & s_{4,\mathrm{I}}[1] \\
11111111 & s_{4,\mathrm{Q}}[1] \\
00000000 & s_{4,\mathrm{I}}[2] \\
00000000 & s_{4,\mathrm{Q}}[2] \\
11111111 & s_{4,\mathrm{I}}[3] \\
00000000 & s_{4,\mathrm{Q}}[3] \\
11111111 & s_{4,\mathrm{I}}[4] \\
11111111 & s_{4,\mathrm{Q}}[4]
\end{array}\right\}$$

Figure 5.2 Example interleaved digital baseband signals for all four transmitters.

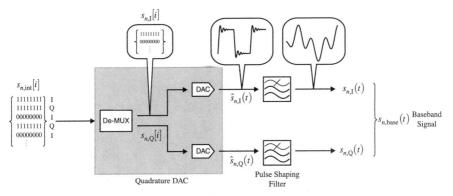

Figure 5.3 Showing conversion from digital baseband chips $s_{n,\text{int}}(i)$ to baseband signal $s_{n,\text{base}}(t)$.

in-phase and quadrature components of the nth baseband signal $s_{n,\text{base}}(t)$, where

$$s_{n,\text{base}}(t) = s_{n,\text{I}}(t) + js_{n,\text{Q}}(t). \tag{5.5}$$

5. Modulate Baseband Signals

The final step in the transmit chain is to convert the baseband signals to a form suitable for transmission over the air. A *quadrature modulator* is used to modulate the amplitude and phase of a carrier according to $s_{n,\text{base}}(t)$. A block diagram of the quadrature modulator is shown in Fig. 5.4. The frequency spectrum of the nth output signal $s_n(t)$ is centered at f_c, which in turn is determined by a *local oscillator* (LO). The same carrier is used at all M_{Tx} transmit chains. The signal $s_n(t)$ is amplified and broadcast across a channel to the receiver.

5.2.2 Reception

Reception involves the following key steps:

1. Demodulate RF Signals

The received signal at each receive-antenna, $r_m(t)$, is demodulated using a *quadrature demodulator*, where $m \in \{1, \ldots, 4\}$. Figure 5.5 shows a block diagram of the quadrature demodulator. The mth demodulator mixes a carrier from a LO at the receiver with $r_m(t)$. The demodulation produces undesired higher-frequency images (107). Furthermore, these images lie outside the first Nyquist zone, and their energy will be aliased back onto the first Nyquist zone after sampling. Therefore, the in-phase and quadrature branches are filtered after demodulation using a passive *anti-aliasing* filter. The cutoff frequency of each filter is determined by the receiver sample rate

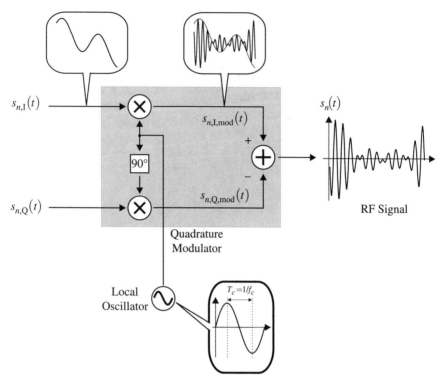

Figure 5.4 Quadrature modulator block diagram for transmitter *n*.

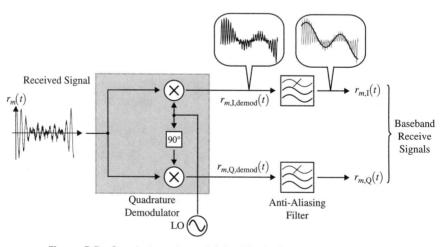

Figure 5.5 Quadrature demodulator block diagram for receiver *m*.

$f_{\text{Rx}} = 1/T_{\text{Rx}}$. The results, after filtering, are the baseband quadrature signals $r_{m,\text{I}}(t)$ and $r_{m,\text{Q}}(t)$.

2. Sample Baseband Signals

Next, two quadrature analog-to-digital converters (ADCs), one in each branch, convert $r_{m,\text{I}}(t)$ and $r_{m,\text{Q}}(t)$ to the digital domain. This is illustrated in Fig. 5.6. Each branch is sampled by an ADC, which yields the baseband receive samples $r_{m,\text{I}}[i]$ and $r_{m,\text{Q}}[i]$. The sample period is determined by a sample clock, whose period is set to T_{Rx}. The WMSDR receiver represents each sample as a 10-bit number. The samples from each branch are interleaved by a multiplexer to yield $r_{m,\text{int}}[i]$.

3. Record Samples to Disk, and Process the Data Offline

During most applications, the baseband samples $r_{m,\text{int}}[i]$ are most often recorded to the receiver PC's hard disk for later analysis. Any number of digital signal processing algorithms can be used to analyze the recorded data. For the purposes of channel sounding, digital matched filtering is used to compute a calibrated estimate of the channel. Digital matched filtering mitigates the nonideal effects of all filters in the transmit and receive chains, such as the anti-aliasing and pulse-shaping filters above. We discuss this further in Section 5.4.3.

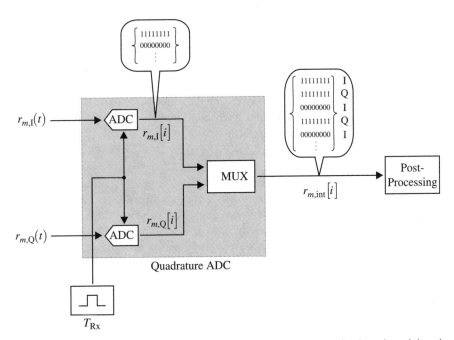

Figure 5.6 Quadrature ADC stage showing conversion from receive baseband signals $r_{m,\text{I}}(t)$ and $r_{m,\text{Q}}(t)$ to the interleaved digital baseband signal $r_{m,\text{int}}(i)$.

5.2.3 Timing and Carrier Offsets

In the case of the WMSDR, the receiver is set so that the oversampling rate at the receiver $R_{over} = T_{Rx}/T_{Tx}$ is ≥ 2. This relaxes the requirements on the anti-aliasing filter and lowers the noise floor in the digital domain by limiting the energy aliased back into the first Nyquist zone.

In addition to oversampling, the receiver sample clock period is not exactly synchronized to its counterpart in the transmitter. That is, T_{Tx} is not always an integer multiple of T_{Rx}, and there is most often a phase difference between both clocks. This leads to *timing offset error. Timing recovery algorithms* are used to recover $s_n[i]$ from $r_{m,int}[i]$. In Section 5.7.1, we discuss two different timing recovery algorithms.

In addition to timing offset error, the LO at the transmitter is not exactly synchronized to its counterpart at the receiver. This leads to differences in the carrier frequency and the carrier phase at both link ends. We refer to this mismatch as *carrier offset error. Carrier recovery algorithms* are used to recover the state of the transmit LO from $r_{m,int}[i]$. We present one such algorithm in Section 5.7.2.

5.3 NARROWBAND CHANNEL SOUNDING

In Chapter 2, we discussed how physical wireless channels can be approximated as linear filters; consequently, the physical channel's impulse response completely characterizes the physical channel. The problem of characterizing a physical channel thus becomes one of estimating its impulse response. This is called *channel sounding.* We refer to an apparatus built specifically to measure the impulse response of a channel as a *channel sounder*, or simply a *sounder*. The channel sounding method used depends on the type of channel. Channel sounding methods can be loosely grouped into two categories: *narrowband* and *wideband channel sounding*. The following introduces several channel sounding methods.

5.3.1 Periodic Pulse Sounding

Most often, channel sounders consist of a separate transmitter and receiver, as shown in Fig. 5.7. Consider the case where each is equipped with a single antenna. The transmitter sends a signal whose frequency content is suitable to infer the frequency response of the channel. For example, sounding a narrowband channel involves the use of a narrowband signal at the transmitter. For a wideband channel, a wideband signal is required. The key to channel sounding is that the transmitted signal is known at the receiver. In this way, the impulse response can be estimated by comparing the transmitted and received signals.

To remove the effect of the channel sounder itself on the measurement, most channel sounders undergo some form of *calibration* before each measurement. If the channel sounder is properly calibrated, any differences between the transmitted and received signals are assumed to be due to the channel.

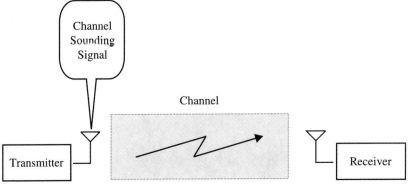

Channel Transfer Function $\rightleftharpoons$ Impulse Response

$$H(f) \rightleftharpoons h(t)$$

Figure 5.7   Typical channel sounding setup.

As the namesake would imply, estimating a channel's *impulse response* involves transmitting an *impulse*, and measuring the *response* at the receiver. This represents the most basic form of channel sounding. The transmitted signal takes the form

$$s_{\text{pps}}(t) = \sum_i \delta(t - iT_{\text{pps}}), \tag{5.6}$$

where T_{pps} is the *pulse period*, and $\delta(t)$ is the delta function. The pulse period is chosen such that the time between pulses is larger than the excess delay of the channel but shorter than the coherence time. This is an example of a *periodic pulse sounder*.

To illustrate this, consider a single period for which $s_{\text{pps}}(t) = \delta(t)$. From linear system theory we know that $\delta(t)$ has a constant frequency spectrum; that is,

$$\delta(t) \rightleftharpoons 1 \quad \text{for all } f. \tag{5.7}$$

At the receiver, we record a frequency-distorted version of $\delta(t)$. This is illustrated in Fig. 5.8. It is easy to infer the channel transfer function $H(f)$ from the received spectrum and thus the complete impulse response of the channel. This makes the periodic pulse sounder useful for wideband channel sounding. For a single period, the received signal $r_{\text{pps}}(t)$ is equal to the channel impulse response $h(t)$; that is, $h(t) = r_{\text{pps}}(t)$ for $iT_{\text{pps}} \leq t < (i+1)T_{\text{pps}}$.

Periodic pulse sounders are difficult to implement in practice. All practical RF components, such as antennae, amplifiers, mixers, and so forth, have a finite bandwidth. Furthermore, it is impossible to generate an ideal impulse at the transmitter. Impulsive signals are also difficult to amplify due to nonlinearities inherent in any RF amplifier. It is also difficult to calibrate a channel sounder over large bandwidths. Most often, we are only interested in a relatively small range of frequencies. Thus, a more efficient design can be implemented by limiting the sounder bandwidth.

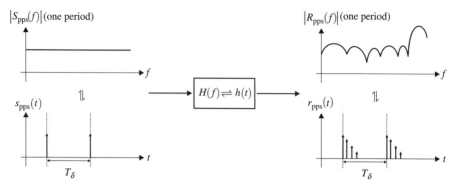

Figure 5.8   Illustration of periodic pulse sounding in the time and frequency domains.

5.3.2 Narrowband Single-Input, Single-Output Channel Sounding

Consider the most specific case, in which we only wish to measure the impulse response of a single narrowband channel. In this case, we consider the response of the channel for a single frequency. By the definition of a narrowband channel, the measured response is constant at all frequencies of interest. From linear system theory, a sinusoid has the following Fourier property:

$$x(t) = \sin(2\pi f_0 t) \rightleftharpoons X(f) = \frac{1}{2j}[\delta(f - f_0) - \delta(f + f_0)]. \qquad (5.8)$$

For $f \geq 0$, the amplitude of the spectrum is

$$|X(f)| = \frac{1}{2}\delta(f - f_0), \qquad (5.9)$$

which only occupies a single frequency f_0. Thus, a sinusoid is the ideal signal to sound the narrowband channel. An ideal narrowband sounding signal $s_{\sin}(t)$ is of the form

$$s_{\sin}(t) = \sin(2\pi f_0 t), \qquad (5.10)$$

where f_0 is the *center frequency*, assumed to be within the bandwidth of interest. At the receiver, we see a distorted version of $s_{\sin}(t)$. This process is illustrated in Fig. 5.9.

Because the channel is assumed to be linear, only the amplitude and phase of the original sinusoid are affected by the channel. Thus, the received signal is of the form

$$r_{\sin}(t) = A\sin(2\pi f_0 t + \vartheta), \qquad (5.11)$$

where A is the amplitude, and ϑ is the phase of the received signal. Most wireless channels are *passive* in that they attenuate, rather than amplify, the transmitted signal. This means that, most often, $A < 1$. Given that the transmitted signal is normalized to have

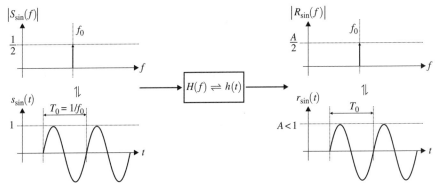

Figure 5.9 Narrowband SISO channel sounding.

unit amplitude and zero phase, the frequency response of the channel is given by the amplitude and phase of $r_{\sin}(t)$; that is,

$$H_{\sin}(f) = A\angle\vartheta. \tag{5.12}$$

Note that the estimate is only valid over the frequency of interest, that is, for $f_0 - W/2 \le f < f_0 + W/2$, where W is the bandwidth of interest in Hz. Note also that $H_{\sin}(f) = H_{\sin}$ does not depend on f, and its time-dual can be written as

$$H_{\sin} \rightleftharpoons h_{\sin}, \tag{5.13}$$

where $h_{\sin}$ is the channel complex gain, and does not depend on t.

5.3.3 Narrowband MIMO Channel Sounding

The narrowband SISO sounding technique can be extended to the narrowband MIMO channel (26). Consider the $M_{Rx} \times M_{Tx}$ case. We are interested in estimating the $M_{Rx} \times M_{Tx}$ complex gains, without having to resort to de-correlation methods to separate the M_{Tx} transmitted signals at the receiver.

To this end, consider the setup shown in Fig. 5.10. A tone is broadcast on each transmit-antenna. The fundamental frequency for each tone is f_n for $n \in \{1, \ldots, M_{Tx}\}$. That is, the signal at the nth transmit-antenna is

$$s_{n,\text{NB}}(t) = \sin(2\pi f_n). \tag{5.14}$$

Each tone, from one antenna to the next, is separated by Δf Hz. The bandwidth of interest W is the range of frequencies covered by all $s_{n,\text{NB}}(t)$; that is, $W = f_1 - f_{M_{Tx}}$. For narrowband measurements, W is chosen to be much smaller than the coherence bandwidth of the channel. Therefore, frequency-selective fading can be ignored within the bandwidth of interest.

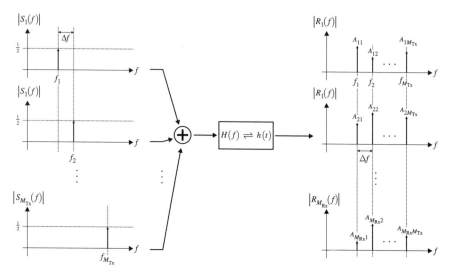

Figure 5.10 Narrowband MIMO channel sounding in the frequency domain.

At the receiver, each antenna captures a distorted version of the sum of all transmitted signals. Specifically, the received signal at antenna m is given by

$$r_{m,\mathrm{NB}}(t) = \sum_{n=1}^{M_{\mathrm{Tx}}} A_{mn} \sin(2\pi f_n t + \vartheta_{mn}), \tag{5.15}$$

where A_{mn} is the amplitude and ϑ_{mn} the phase of the tone from transmit-antenna n as seen by receive-antenna m. Because the tones are separated in frequency by Δf, it is possible to compute the channel gain at each tone for each receive-antenna. Using the same argument as for the narrowband SISO case, the complex gain can be expressed as

$$h_{mn} = A_{mn}\angle\vartheta_{mn}. \tag{5.16}$$

Thus, given all $r_{m,\mathrm{NB}}(t)$ for $m \in \{1, \dots, M_{\mathrm{Rx}}\}$, all possible h_{mn} can be computed and mapped to **H**.

5.4 WIDEBAND SOUNDING: CORRELATIVE SOUNDING

In general, wideband channel sounding is considerably more complex than its narrowband counterpart. Many wideband channel sounders use a *pseudo-noise sequence* (PN sequence) to probe the channel. A PN sequence is a finite-length binary sequence that shares some properties of noise. There are many different types of PN sequences, each with its own autocorrelation and cross-correlation properties. One important requirement is that the PN sequence has a relatively constant, or *flat* spectrum over W.

In addition to its spectrum, the autocorrelation and cross-correlation properties of the PN sequence are of interest, especially in MIMO applications. In PN sequence channel sounding, the receiver computes the *correlation* between the transmitted PN sequence and the received signal to determine the channel impulse response. In this instance, we speak of a *correlative channel sounder*. The following describes the theory behind correlative channel sounding.

Consider the correlative channel sounder setup shown in Fig. 5.11. The transmitted signal consists of a finite sequence of pulses. Specifically, let $s_{PN}(t)$ represent the signal at the transmitter,

$$s_{PN}(t) = \sum_{i=0}^{L_{PN}-1} a_i p(t - iT_{PN}), \tag{5.17}$$

where $p(t)$ is a function defining the pulse shape, L_{PN} is the PN sequence length, T_{PN} is the PN sequence chip period, and a_i is the ith pulse amplitude. The pulse amplitude $a_i \in \{-1, 1\}$ is mapped directly from the binary PN sequence. The *PN-sequence period* is given by $T_{L,PN} = T_{PN}L_{PN}$.

Consider the specular time-invariant channel model given by (108)

$$h(t) = \sum_{d=0}^{D-1} h[d]\delta(t - \tau_d), \tag{5.18}$$

where D is the total number of arrivals, $h[d]$ is the complex channel gain at delay d, and τ_d is the time delay of the dth arrival. Given a linear channel, the channel output is computed as

$$y(t) = h(t) * s_{PN}(t)$$

$$= \sum_{d=0}^{D-1} h[d] \cdot s_{PN}(t - \tau_d) \tag{5.19}$$

$$= \sum_{d=0}^{D-1}\sum_{i=0}^{N-1} h[d] \cdot a_i p(t - \tau_d - iT_{PN}),$$

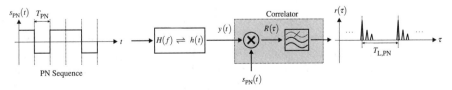

Figure 5.11 PN-sequence correlative channel sounding.

where $y(t)$ is the channel output. The channel output is correlated with the transmitted signal to estimate the channel impulse response, yielding

$$R(\tau) = \frac{1}{T_{L,PN}} \int_{-T_{L,PN}/2}^{T_{L,PN}/2} y(t)s_{PN}(t - \tau)dt$$

$$= \sum_{d=0}^{D-1} \sum_{i=0}^{L_{PN}-1} \sum_{j=0}^{L_{PN}-1} h[d]a_i a_j \frac{1}{T_{L,PN}} \int_{-T_{L,PN}/2}^{T_{L,PN}/2} p(t - \tau_d - iT_{PN})p(t - jT_{PN} - \tau)dt.$$

$$(5.20)$$

Let the result of the convolution integral in (5.20) be defined by

$$R_c(\tau) = \frac{1}{T_{L,PN}} \int_{-T_{L,PN}/2}^{T_{L,PN}/2} p(t - dT_\delta - iT_{PN})p(t - jT_{PN} - \tau)dt. \qquad (5.21)$$

Combining (5.20) and (5.21), the correlator output $R(\tau)$ can be expressed as

$$R(\tau) = \sum_{d=0}^{D-1} \sum_{i=0}^{N-1} \sum_{j=0}^{N-1} h[d]a_i a_j \cdot R_c(\tau). \qquad (5.22)$$

The correlator output consists of the channel impulse response $h[d]$, an amplitude product term $a_i a_j$, and the autocorrelation term $R_c(\tau)$, all of which are dependent on the PN sequence. Thus, the selection of $s_{PN}(t)$ becomes important, especially in MIMO channels where more than one sequence is used simultaneously. It is possible to design the PN sequence to obtain an unbiased estimate of $h[d]$, as described next.

5.4.1 ML Sequences

Maximal-length sequences, or *ML sequences*, are a special type of PN sequence with cross-correlative and autocorrelative properties useful to wideband channel sounding. ML sequences are used, for example, in multiuser spread-spectrum communication systems to encrypt different users at their respective transmitters, making it possible to separate users at any given receiver (109). The traditional definition holds that ML sequences are binary; that is, each symbol in the sequence is chosen from one of two possible symbols. However, other examples of ML sequences exist where each level is chosen from one of $L_s = 3$ or 5 possible levels (110).

Traditionally, binary ML sequences are generated using a combination of a shift register of length m_{ML} and modulo-2 adders. If the shift register is preset to any nonzero state, it produces a sequence of length $L_{ML} = 2^{m_{ML}} - 1$ before repeating itself. Thus, *one unique sequence* can be generated for a given m_{ML}. The shift register is clocked such that it produces a new symbol at the rising edge of each clock. The

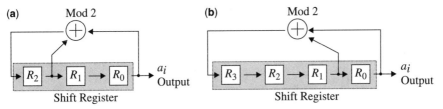

Figure 5.12 ML-sequence generator for (a) $m_{\text{ML}} = 3$ and (b) $m_{\text{ML}} = 4$.

block diagrams for the cases where $m_{\text{ML}} = 3$ and 4 are shown in Fig. 5.12. All binary ML sequences exhibit the following properties (106):

1. The number of 1s in any given sequence is one more than the number of 0s. Thus, ML sequences are said to be *balanced*.
2. In an ML sequence, there are subsequences consisting of repeated 1s or 0s. Any subsequence of repeated values is called a *run*. For any ML sequence, half of the runs are of length 1, a quarter of the runs are of length 2, and so forth, as long as the fractions represent a possible number of runs. This is called the *run property*.

Properties 1 and 2 ensure that ML sequences have excellent autocorrelation and cross-correlation properties.

Given the prevalence of digital hardware, we often see the ML sequence expressed using discrete variables. Let $b_i \in \{0, 1\}$ represent a binary ML sequence of length L_{ML}. Let $p[i] \in \{-1, 1\}$ represent the antipodal signal, computed as

$$p[i] = 2b_i - 1 \tag{5.23}$$

for $i \in \{1, \ldots, L_{\text{ML}}\}$. The autocorrelation function $R_{\text{ML}}[d]$ is periodic, with period L_{ML}. For one period, the (normalized) autocorrelation function is (109)

$$R_{\text{ML}}[d] = \frac{1}{L_{\text{ML}}} \sum_{i=1}^{L_{\text{ML}}} p[i]p[i+d]$$
$$= \begin{cases} 1 & \text{for } d = 0 \\ -1/L_{\text{ML}} & \text{for } 1 \leq d \leq L_{\text{ML}}. \end{cases} \tag{5.24}$$

The autocorrelation function is plotted in Fig. 5.13 for the case where $L_{\text{ML}} = 15$.

The binary ML sequence is similar to a *random binary wave* in that they both have similar autocorrelation profiles. The most notable difference is that an ML sequence is *finite*, and thus the autocorrelation is periodic, with period L_{ML}. This means that the power spectrum for a random binary wave is also similar to that of an ML sequence, except that the ML sequence spectrum is a *sampled* version of the binary wave, which follows from linear system theory. The sample period corresponds with the inverse of the ML-sequence period L_{ML}.

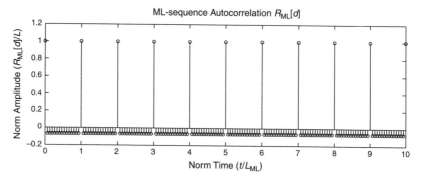

Figure 5.13 The discrete ML sequence autocorrelation function $R_{ML}(d)$ for the case when $L_{ML} = 15$.

5.4.2 Cross-correlation Using the FFT

Consider the setup shown in Fig. 5.14, where all signals are in discrete time. This setup illustrates a method by which we can obtain the autocorrelation of an ML sequence in the digital domain. This technique relies on the *fast-Fourier transform* (FFT) and the inverse-FFT (iFFT). First, the transmitter generates a discrete binary ML sequence $p[i]$. Here, it is assumed that $p[i]$ is a finite length sequence of length L_{ML}. To compute $R_p[d]$, the receiver first computes the spectrum of $p[i]$ via the FFT; that is, $p[i] \underset{\text{FFT}}{\rightleftharpoons} P[f]$. $P[f]$ is then multiplied with its complex conjugate, $P^*[f]$. With respect to Fig. 5.14, at the output of the multiplier, this yields

$$Y[f] = P[f] \cdot P^*[f]. \tag{5.25}$$

The output is converted back to the time domain using the iFFT. The result is given by (61)

$$y[d] = \frac{1}{L_{ML}} \text{iFFT}\{P[f] \cdot P^*[f]\}$$

$$= \frac{1}{L_{ML}} \sum_{f=1}^{L_{ML}} p[f] p^*[f - d]\Big|_{\text{mod } L_{ML}}, \tag{5.26}$$

$$= R_p[d]$$

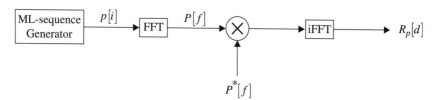

Figure 5.14 Computing the autocorrelation of a discrete sequence $p(i)$ using the digital matched filtering technique.

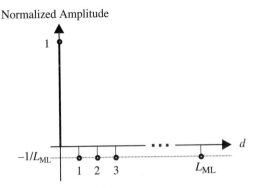

Figure 5.15 Autocorrelation function $R_p(d)$ for a binary ML sequence $p(i)$ of length L_{ML}.

where the autocorrelation function $R_p[d]$ is also a finite discrete function of length L_{ML}, and

$$R_p[d] = \begin{cases} 1 & \text{for } d = 0 \\ -1/L_{\text{ML}} & \text{for } 1 \le d \le L_{\text{ML}}. \end{cases} \tag{5.27}$$

Comparing (5.27) and (5.24), we see that $R_p[d]$ is similar to one period of $R_{\text{ML}}[d]$. $R_p[d]$ is plotted in Fig. 5.15 for the general case.

5.4.3 Digital Matched Filters

Digital matched filtering was first introduced by Fannin et al. in Ref. 111. In the following, we consider the case where we wish to estimate the impulse response of a channel using ML sequences at the transmitter and the FFT technique described above to perform cross-correlation at the receiver. Digital matched filtering has the added benefit that it calibrates out the frequency-selective (nonideal) effects of the sounder itself. This gives a more accurate estimate of the channel impulse response. In the following, we discuss wideband sounder calibration, digital matched filtering, and its generalization to the MIMO case.

5.4.3.1 Wideband Sounder Calibration Ideal wideband channel sounders have no frequency-selective characteristics within the bandwidth of interest. Put differently, both the transmitter and receiver have a *flat frequency response* over the entire sounder bandwidth. This ensures that the sounder does not bias the channel estimate. This is very difficult to achieve in practice. Most wideband channel sounders suffer from some form of frequency selectivity at the transmitter and receiver. This is most notably due to the nonideal transfer functions of the *transmit-filters* and *receive-filters*.

Consider the simplified transmitter/receiver architecture shown in Fig. 5.16. Both the transmitter and receiver are typically equipped with many filters, each with a nonideal transfer function. The purpose of each filter is briefly described here. At the transmitter, the *reconstruction filter* is tuned to suppress images produced by the

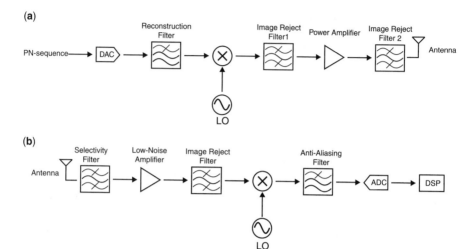

Figure 5.16 A simplified transmitter–receiver block diagram for a typical digital radio highlighting the location of different filters in the (a) transmit and (b) receive chains and their purpose.

digital-to-analog conversion. The first *image reject filter* reduces images produced by the mixer before the signal is amplified. A second image reject filter supresses images caused by any nonlinearities in the power amplifier. At the receiver, a *selectivity filter* supresses the out-of-band noise power and reduces interference from other sources. The image reject filter after the low noise amplifier (LNA) helps reduce images produced by nonlinearities in the LNA while improving mixer performance. Finally, the anti-aliasing filter reduces any signal energy outside of the first Nyquist zone.

It is difficult to design all the above filters such that their frequency response is flat across the entire sounder bandwidth. In addition, the components between filters can also exhibit some frequency selectivity. We need some way to remove or mitigate any frequency selectivity across the sounder bandwidth. This is part of the *calibration* process.

5.4.3.2 Measuring a Matched Filter

Digital matched filtering is a relatively simple way of removing most frequency selectivity in the sounder electronics via software. It involves first measuring the transfer function of the sounder when the transmitter is connected directly to the receiver and then "dividing" this data out of subsequent measurements.

Consider the setup shown in Fig. 5.17, where the transmitter is connected directly to the receiver. In this case, we say the transmitter and receiver are connected *back-to-back*.

Here, $L_{Tx}[f]$ and $L_{Rx}[f]$ represent the total transfer function of all filters at the transmitter and receiver, respectively. Given that the receiver knows the ML-sequence spectrum $P[f]$, the *back-to-back received signal* $N[f]$ can be used to

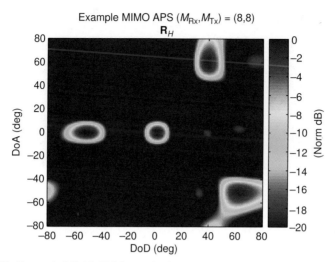

Figure 2.15 Example MIMO APS for an 8 × 8 showing four distinct resolvable scatterers.

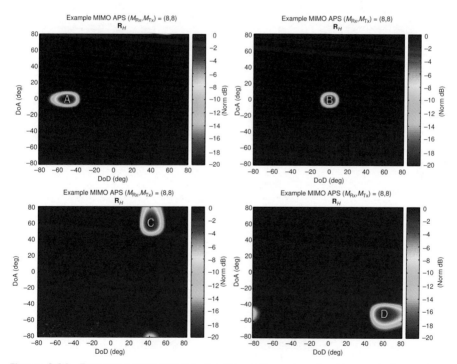

Figure 2.16 Example MIMO APS showing the scatterer described by each eigenvalue. Note that the complete MIMO APS, formed by summing the contribution from all eigenvalues, resembles the MIMO APS shown in Fig. 2.15, which was formed using four *correlated* scatterers.

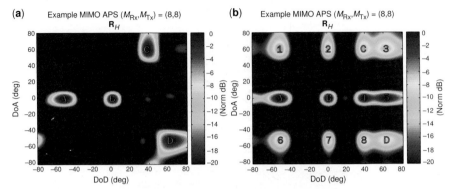

Figure 3.2 The effects of the separability assumption on the MIMO APS. (a) MIMO APS of the true channel, showing four resolvable scatterers labeled A–D. (b) MIMO APS using the Kronecker model, showing artifact paths labeled 1–8.

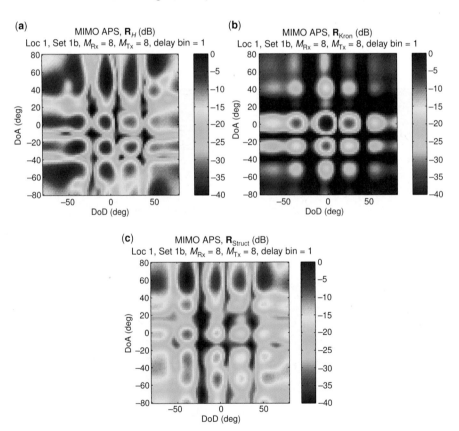

Figure 6.8 MIMO APS for BYU data, location 1, set 1b. (a) True MIMO APS computed from $\mathbf{R}_{\text{WB},H(d)H(d)}$. (b) MIMO APS computed from $\mathbf{R}_{\text{WB},H(d)H(d),\text{Kron}}$. (c) MIMO APS computed from $\mathbf{R}_{\text{WB},H(d)H(d),\text{struct}}$.

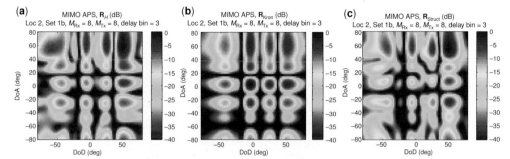

Figure 6.9 MIMO APS for BYU data, location 2, set 1b, delay 3. (a) True MIMO APS computed from $\mathbf{R}_{\mathrm{WB},H(d)H(d)}$. (b) MIMO APS computed from $\mathbf{R}_{\mathrm{WB},H(d)H(d),\mathrm{Kron}}$. (c) MIMO APS computed from $\mathbf{R}_{\mathrm{WB},H(d)H(d),\mathrm{struct}}$.

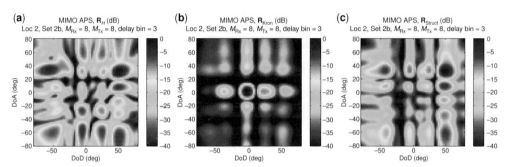

Figure 6.10 MIMO APS for BYU data, location 2, set 2b, delay 3. (a) True MIMO APS computed from $\mathbf{R}_{\mathrm{WB},H(d)H(d)}$. (b) MIMO APS computed from $\mathbf{R}_{\mathrm{WB},H(d)H(d),\mathrm{Kron}}$. (c) MIMO APS computed from $\mathbf{R}_{\mathrm{WB},H(d)H(d),\mathrm{struct}}$.

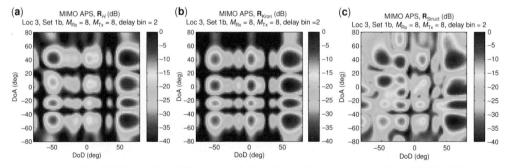

Figure 6.11 MIMO APS for BYU data, location 3, set 1b, delay 2. (a) True MIMO APS computed from $\mathbf{R}_{\mathrm{WB},H(d)H(d)}$. (b) MIMO APS computed from $\mathbf{R}_{\mathrm{WB},H(d)H(d),\mathrm{Kron}}$. (c) MIMO APS computed from $\mathbf{R}_{\mathrm{WB},H(d)H(d),\mathrm{struct}}$.

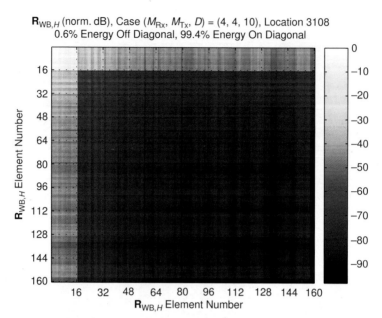

Figure 6.12 The wideband correlation matrix $\mathbf{R}_{WB,H}$ plotted for the WMSDR data for location 3108.

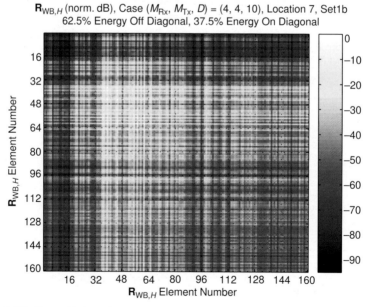

Figure 6.13 The wideband correlation matrix $\mathbf{R}_{WB,H}$ plotted for the BYU data for location 7, set 1b.

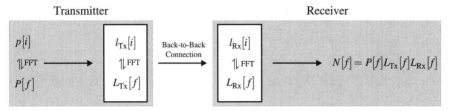

Figure 5.17 Showing the back-to-back connection $N(f)$.

compute the transfer function of the sounder. Specifically, at the receiver, the product

$$N[f]N^*[f] = |L_{Tx}[f]L_{Rx}[f]|^2|P[f]|^2 \qquad (5.28)$$

can be computed. The total transfer function of the sounder is given by

$$|L_{Tx}[f]L_{Rx}[f]|^2 = \frac{N[f]N^*[f]}{|P[f]|^2}. \qquad (5.29)$$

The *matched filter*, denoted by $M[f]$, is defined to be

$$M[f] \triangleq \frac{N^*[f]}{|L_{Tx}[f]L_{Rx}[f]|^2}. \qquad (5.30)$$

Now consider the case illustrated in Fig. 5.18. This setup differs from that in Fig. 5.17 in that $N[f]$ is now multiplied by the matched filter $M[f]$. By doing this, the autocorrelation $R_p[d]$ can be computed by removing the effects of $L_{Tx}[f]$ and $L_{Tx}[f]$. Analytically, the *calibrated* back-to-back received signal $N_c[f]$ is computed as

$$N_c[f] = P[f]L_{Tx}[f]L_{Rx}[f] \cdot M[f]. \qquad (5.31)$$

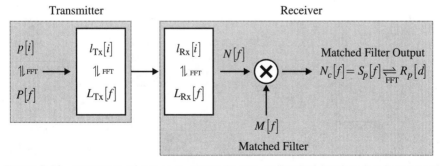

Figure 5.18 Illustration of how we compute the calibrated back-to-back received signal $N_c(f)$.

Substituting (5.30) into (5.31), we get

$$N_c[f] = P[f]L_{\text{Tx}}[f]L_{\text{Rx}}[f] \cdot \frac{N^*[f]}{|L_{\text{Tx}}[f]L_{\text{Rx}}[f]|^2}. \tag{5.32}$$

From Fig. 5.17, $N[f] = P[f]L_{\text{Tx}}[f]L_{\text{Rx}}[f]$. Substituting this equation into (5.32),

$$N_c[f] = P[f]P^*[f] \cdot \frac{|L_{\text{Tx}}[f]L_{\text{Rx}}[f]|^2}{|L_{\text{Tx}}[f]L_{\text{Rx}}[f]|^2} \tag{5.33}$$

$$= P[f]P^*[f].$$

The autocorrelation $R_p[d]$ is computed by taking the iFFT of $N_c[f]$, that is,

$$R_p[d] = \frac{1}{L_m} \text{iFFT}\{P[f]P^*[f]\}$$

$$= \begin{cases} 1 & \text{for } d = 0 \\ -1/L_m & \text{for } 1 \le d \le L_m, \end{cases} \tag{5.34}$$

which is identical to the correlator output of (5.27). Thus, the above shows how the digital matched filtering technique can theoretically be used to mitigate any nonideal frequency effects of the sounder.

5.4.3.3 Matched Filter Channel Estimation

Here, we discuss how digital matched filtering can be used to obtain an unbiased estimate of the channel transfer function. Consider the sounder setup shown in Fig. 5.19. The setup shown in Fig. 5.19 is similar to that of Fig. 5.18, with the important exception that the channel $H[f]$ exists between the transmitter and receiver. Here the channel is assumed to be time-invariant but wideband. The received spectrum $N_H[f]$ is

$$N_H[f] = P[f]P^*[f] \cdot H[f]. \tag{5.35}$$

Note that the ML-sequence amplitude spectrum $|P[f]|^2 = 1$ for all f, and thus the received spectrum is equal to the Fourier transform of the impulse response; that is,

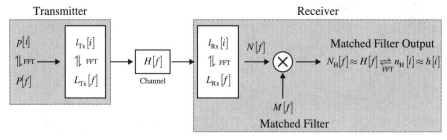

Figure 5.19 Use of digital matched filtering to obtain a calibrated estimate of the channel transfer function $H(f)$.

$N_H[f] = H[f]$. The sampled impulse response $h[i]$ is estimated by taking the iFFT of $N_H[f]$,

$$
\begin{aligned}
n_H[i] &= \text{iFFT}\{P[f]P^*[f]\} * h[i] \\
&= R_p[i] * h[i].
\end{aligned}
\tag{5.36}
$$

For long sequences, $R_p[i]$ can be considered to be impulsive, in which case we may set $n_H[i] \approx h[i]$ (112). Thus, the calibrated estimate of the wideband channel impulse response can be obtained using the technique of digital matched filtering.

5.4.3.4   *MIMO Digital Matched Filtering*

As previously mentioned, ML sequences have excellent cross-correlation properties. This makes them ideal for use in wideband multiple antenna correlative channel sounders, such as the WMSDR, where we can transmit more than one ML sequence simultaneously. In this section, we discuss the extension of matched filter channel sounding to the MIMO case. Digital matched filtering is used to obtain the channel estimates from the WMSDR data in Chapter 6.

Figure 5.20 illustrates the back-to-back connection between receiver m and transmitter n. Each transmit antenna broadcasts a unique ML sequence of length L_{ML}, whose spectra are given by $P_n[f]$. To produce a unique sequence at each transmitter, the $p_n[i]$ are shifted in time, that is, if $p_1[i] = p[i]$, then $p_n[i] = p[i + \Delta d_n]|_{\text{mod } L_{ML}}$, where Δd_n is some arbitrary delay, and $p_n[i]$ is the ML sequence at transmit antenna n. Possible values for Δd_n include $\{1, \ldots, L_{ML} - 1\}$. In practice, the delay must be chosen such that the time difference between sequences $\Delta d_{n+1} - \Delta d_n$ exceeds the maximum resolvable delay of the channel. This will be explained in the following.

For the case shown in Fig. 5.20, the back-to-back received signal spectrum $N_{mn}[f]$ is

$$
N_{mn}[f] = P_n[f]L_{\text{Tx},n}[f]L_{\text{Rx},m}[f].
\tag{5.37}
$$

For the $M_{\text{Rx}} \times M_{\text{Tx}}$ MIMO case, $M_{\text{Rx}} \cdot M_{\text{Tx}}$ matched filter coefficients $M_{mn}[f]$ must be recorded, one for every possible antenna pair. The (m, n)th matched filter coefficient is defined as

$$
M_{mn}[f] = \frac{N_{mn}^*[f]}{|L_{\text{Tx},n}[f]L_{\text{Rx},m}[f]|^2},
\tag{5.38}
$$

where $N_{mn}[f]$ is the received signal from the back-to-back connection between transmit antenna n and receive antenna m, $L_{\text{Tx},n}[f]$ is the combined transfer function

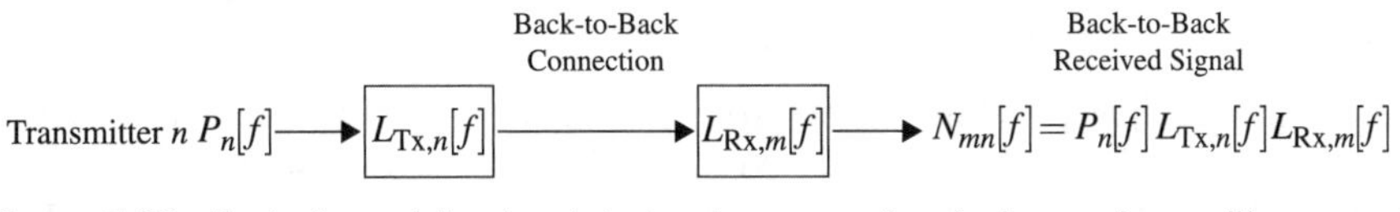

Figure 5.20   Illustration of the back-to-back connection between transmitter n and receiver m.

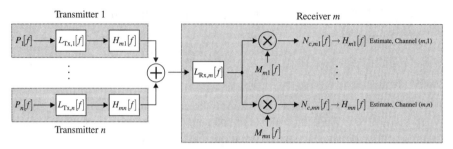

Figure 5.21 Illustration of the received signal at the mth receive-antenna, after digital matched filtering.

of the nth transmit chain, and $L_{\text{Rx},m}[f]$ is the combined transfer function of the mth receive chain.

Figure 5.21 shows the block diagram of the signal received by receiver m, which is composed of all M_{Tx} transmitted signals, distorted by the time-invariant channel $H_{mn}[f]$. The calibrated signal at receiver m is thus

$$
\begin{aligned}
N_{c,mn}[f] &= P_1[f]L_{\text{Tx},1}[f]H_{m1}[f]L_{\text{Rx},m}[f]M_{mn}[f] + \cdots \\
&\quad + P_n[f]L_{\text{Tx},n}[f]H_{mn}[f]L_{\text{Rx},m}[f]M_{mn}[f] + \cdots \\
&\quad + P_{M_{\text{Tx}}}[f]L_{\text{Tx},M_{\text{Tx}}}[f]H_{mM_{\text{Tx}}}[f]L_{\text{Rx},m}[f]M_{mn}[f] \\
&= H_{mn}[f]P_n[f]P_n^*[f] + \{\text{uncalibrated terms}\},
\end{aligned}
\tag{5.39}
$$

where the *uncalibrated terms* are defined as

$$
\begin{aligned}
\{\text{uncalibrated terms}\} &= H_{m1}[f]P_1[f]P_n^*[f]\frac{L_{\text{Tx},1}[f]L_{\text{Tx},n}^*[f]}{|L_{\text{Tx},n}[f]|^2} + \cdots \\
&\quad + H_{mM_{\text{Tx}}}[f]P_{M_{\text{Tx}}}[f]P_n^*[f]\frac{L_{\text{Tx},M_{\text{Tx}}}[f]L_{\text{Tx},n}^*[f]}{|L_{\text{Tx},n}[f]|^2}.
\end{aligned}
\tag{5.40}
$$

We can estimate $h_{mn}[i]$ by taking the iFFT of $N_{c,mn}[f]$,

$$
n_{c,mn}[i] \approx h_{mn}[i] + \text{iFFT}\{\text{uncalibrated terms}\}.
\tag{5.41}
$$

The signal $n_{c,mn}[i]$ contains an unbiased estimate of $h_{mn}[i]$, as well as uncalibrated estimates for all the other $M_{\text{Rx}}M_{\text{Tx}} - 1$ channels. The key here is that each channel estimate is separated in time. For $h_{mn}[i]$, if we choose the time difference $\Delta d_{n+1} - \Delta d_n$ for all n such that the separation between sequences is greater than the maximum excess delay of all channels, then

$$
n_{c,mn}[i] \approx h_{mn}[i] \quad \text{for } 0 \leq i \leq \Delta d_n - 1,
\tag{5.42}
$$

which is a calibrated estimate of the (m, n)th channel. Consider the 2×1 example shown in Fig. 5.22. Here, $n_{c,11}[i]$ consists of the unbiased estimate $h_{11}[i]$, and the uncalibrated estimate $h_{12}[i]$, separated in time by Δd_2. Thus $n_{c,11}[i] \approx h_{11}[i]$ for $0 \leq i \leq \Delta d_1$. Using the above technique, we can compute all $M_{\text{Rx}}M_{\text{Tx}}$ calibrated channel estimates.

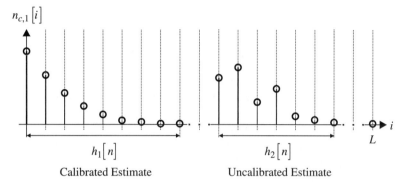

Figure 5.22 Example matched filter output $n_{c,11}(i)$, which contains a calibrated estimate of the impulse response of the first channel $h_{11}(i)$ and a distorted estimate of the second channel $h_{12}(i)$.

Using WMSDR data, Fig. 5.23 shows the resulting PDPs for the first receiver, recorded when the transmitter was wired directly to the receiver. All PDPs were computed using the digital matched filtering technique outlined above and 4095-chip *binary phase-shift keyed* (BPSK) ML sequences.

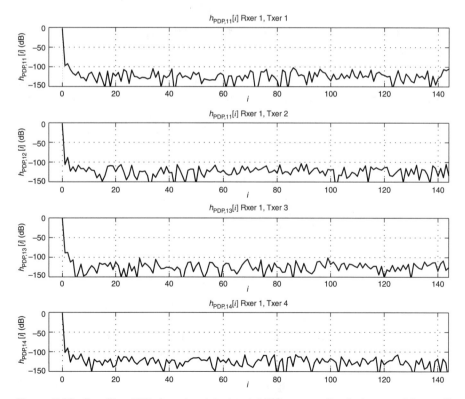

Figure 5.23 Resulting PDPs from back-to-back MISO connection between all transmit antennas and receiver 1.

Let $h_{\mathrm{PDP},mn}[i]$ represent the PDP related to the SISO channel between transmitter n, receiver m, where

$$h_{\mathrm{PDP},mn}[i] = 20 \log |n_{c,mn}[i]|. \tag{5.43}$$

In each PDP, we see that there is a clear peak where the transmitted sequence correlates with itself in the receive data. Furthermore, this correlation is exceptionally sharp; there is only one peak where the two sequences line up, and the magnitude of the next few samples, which we refer to as the *shoulders*, are well below the peak. For example, looking at $h_{\mathrm{PDP},13}[i]$, the transmit sequence lines up with itself at $i = 0$, where $h_{\mathrm{PDP},13}[0] = 0$ dB. The peaks are automatically normalized as a result of the above technique. The shoulders of the peak fall below -85 dB. In addition, the noise floor for the rest of the graph falls well below -100 dB. This compares favorably with results from other sounders, such as the the Brigham Young University (BYU) wideband channel sounder.

The biggest disadvantage of the digital matched filter technique is that it is not applicable to the time-variant wideband channel (i.e., the above analysis will not hold if $H_{mn}[f]$ is also a function of t). This precludes its use in, for example, high-speed mobile channels. In Refs. 113 and 114, Matz et al. extend digital matched filtering to include the time-variant case.

5.5 WIDEBAND SOUNDING: SAMPLED SPECTRUM CHANNEL SOUNDING

A much simpler method of obtaining the impulse response of a time-invariant wideband channel involves sampling the channel transfer function at discrete points in frequency. This channel estimation method has become more popular in recent years mainly due to the increased availability of inexpensive *RF function generators*. The BYU wideband MIMO channel sounder, to be described in Chapter 6, is an example of a sampled spectrum sounder.

Consider the simplified setup shown in Fig. 5.24. The transmitter is equipped with an RF function generator. The RF function generator outputs a *multitone signal*, which consists of T tones, spaced in frequency by Δf. The spectrum at the output of the RF function generator can be expressed as

$$S_{\mathrm{ss}}(f) = \sum_{i=0}^{T-1} \delta(f - i\Delta f), \tag{5.44}$$

where $S_{\mathrm{ss}}(f)$ is the multitone signal, which is illustrated in Fig. 5.25. In this way, the sounding signal can be viewed as T narrowband tones, each sounding a small portion of the channel. If $H(f)$ does not vary appreciably in between each tone, we can reconstruct it from its samples. This is analogous to the sampling theorem but applied to the frequency domain.

At the receiver, the received signal is $R(f) \leftrightharpoons r(t)$. If T is chosen to be a power of 2, the T-point FFT can be computed to obtain a *sampled* version of the channel transfer

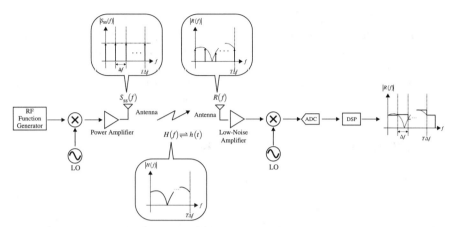

Figure 5.24 Simplified transmitter–receiver setup for a sampled-spectrum wideband channel sounder illustrating the frequency domain signals at different points.

function. By doing this, each *frequency bin* of the discrete spectrum $R[f]$ contains the energy from a single tone of the multitone signal. Each frequency bin can be thought of as the transfer function of the narrowband channel centered at that frequency. That is, $R[f_t]$ can be interpreted as the narrowband channel estimate at $f = f_t$, and therefore $R[f] \approx H[f]$.

It should be noted that although the sampled spectrum topology is used in many MIMO channel sounders such as the MEDAV (115) and BYU wideband channel sounders (116), it is impractical to use a multitone signal with *true* MIMO channel sounders. The reason for this is as follows. Consider the $M_{Rx} \times M_{Tx}$ case. Using a multitone signal at all M_{Tx} antennas implies the use of M_{Tx} signal generators, one for each antenna. In addition, each multitone signal must be designed such that they can be distinguished at the receiver. Given that each tone must occupy approximately the same frequency to sound the same "slice" of bandwidth, it is difficult to achieve this with a multitone signal.

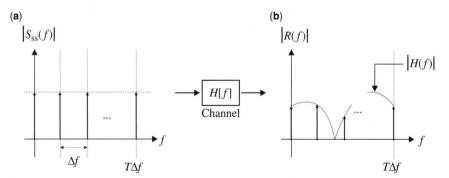

Figure 5.25 (a) An example transmitted multitone signal. (b) An example received multitone signal after transmission through a wideband channel.

Wideband MIMO channel sounders that use the sampled spectrum technique often rely on a *switched-array* architecture to sound one of $M_{Rx} \cdot M_{Tx}$ possible channels at a time. The key is to sound all $M_{Rx} \cdot M_{Tx}$ channels before the channel changes, or within its coherence time. True MIMO channel sounders, such as the WMSDR, sound all $M_{Rx} \cdot M_{Tx}$ possible channels simultaneously. The alternate scheme, switched-array channel sounders, is discussed in more detail in the following section.

5.6 SWITCHED-ARRAY ARCHITECTURES

Switched-array sounders reduce the complexity of the overall system versus a true MIMO sounder but also severely limit its functionality. In contrast with the WMSDR, most switched-array sounders serve a single purpose in that they are only capable of sounding the channel. This section outlines the fundamentals and some of the advantages and disadvantages of switched-array architectures.

Consider the block diagram shown in Fig. 5.26. A switched-array sounder can be thought of as a SISO sounder, equipped with RF switches at both transmit-antenna and receive-antenna arrays. The transmitter distributes an RF signal to one of M_{Tx} antennas. The transmitter dwells on each transmit-antenna for a given time, say $t_{Tx,dwell}$. It takes some finite time to scan the entire array, which we designate $t_{Tx,scan}$. The receiver records information from each of M_{Rx} antennas for $t_{Rx,dwell}$, and it takes $t_{Rx,scan}$ to scan its entire array.

To record a complete MIMO response, the transmitter must scan its entire array within a single $t_{Rx,dwell}$. The transmitter must then rescan its array just after the receiver switches to its next antenna. This process is repeated for all M_{Rx} antennas. Because of the complex timing requirements, it is imperative that both the transmitter and receiver be closely synchronized. Both require an accurate time reference that is typically synchronized before each measurement. Furthermore, the references cannot drift relative to one another over the course of a measurement.

In practice, the impulse response of a channel changes over time. For this reason, it is important to record a complete MIMO response before the channel changes appreciably. This places an upper limit on $t_{Rx,scan}$, and hence on all the other timing specifications. Because not all antennas in the array can be accessed simultaneously, it is impossible to transmit or receive different signals on different antennas simultaneously. This restricts the use of a switched-array architecture to channel sounding only.

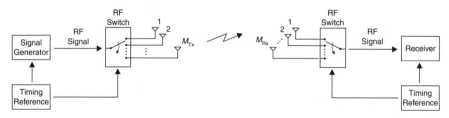

Figure 5.26 The switched-array architecture.

If the purpose is only to sound the channel, then the advantages of switched-array sounders far outweigh their disadvantages. Most notably, the setup involves broadcasting only one RF signal at the transmitter and recording one signal at the receiver. This greatly reduces the hardware complexity. Another advantage of switched sounders is that the maximum size of the transmit-array and receive-array (i.e., the number of elements) is largely a function of the speed of the RF switches and the coherence time of the channel we wish to measure.

Yet another advantage is that, because the receiver only records one signal as opposed to M_{Rx} different signals, switched-array sounders tend to have much larger *effective sounding bandwidths* than true MIMO sounders. The effective sounding bandwidth is usually limited by the maximum sample rate at the receiver. The maximum sample rate, in turn, is often limited by the *maximum data rate* of the receiver.

For example, say the receiver has a maximum data rate of 32 MBytes/s and that the required resolution of each sample is 16-bits. A switched-array receiver would have to record only one stream at 16-bits $= 2$ Bytes/sample, which means we can have a maximum sample rate of

$$\frac{32 \text{ MBytes/s}}{2 \text{ Bytes/sample}} = 16 \text{ MSamples/s}. \tag{5.45}$$

This translates to an effective sounding bandwidth, defined as follows in accordance with Nyquist's sampling theorem:

$$\frac{16 \text{ MSamples/s}}{2} = 8 \text{ MHz}. \tag{5.46}$$

Now consider the MIMO case, where the receiver must record four data streams simultaneously. Also, the receiver is restricted to the same maximum data rate and resolution. This means that our maximum sample rate is now

$$\frac{32 \text{ MBytes/s}}{(2 \cdot 2 \cdot 2 \cdot 2) \text{ Bytes/sample}} = 4 \text{ MSamples/s}, \tag{5.47}$$

and the effective sounding bandwidth is

$$\frac{4 \text{ MSamples/s}}{2} = 2 \text{ MHz}. \tag{5.48}$$

Given that the data rate of the receiver is usually the biggest bottleneck, the above example illustrates one of the biggest advantages of switched-array architecture sounders.

5.7 TIMING AND CARRIER RECOVERY

Consider the digital transmitter-receiver setup shown in Fig. 5.27. At the transmitter, the DAC stage converts digital data to a quadrature analog baseband signal. This signal is mixed with a carrier and broadcast over the air. At the receiver we perform the

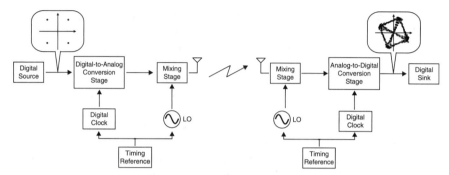

Figure 5.27   Simplified transmitter–receiver setup highlighting sources of error that arise from timing mismatches at both link ends. In this example, the digital timing and LO frequency are phase-locked to a common timing reference.

reverse operation; the received signal is down-converted to baseband, and the resulting signal is digitized. The data conversion stage at the transmitter and receiver is driven by a sample clock, and the mixing stage is driven by an LO. In many cases, both the LO and the sample clock are tied into a common *timing reference*. In this way, the timing reference determines both the *phase* and *frequency* of the sample clock and the phase and frequency of the local oscillator.

To recover the transmitted signal properly, the receiver must sample the incoming signal at the right sample instants. To do this, the receiver must recover the phase and frequency of the transmitter sample clock and local oscillator. We refer to the process of recovering the state of the transmitter sample clock and local oscillator respectively as *timing* and *carrier recovery*. After performing this step, the receiver is said to be *synchronized* to the transmitter.

Synchronization methods for channel sounders fall under four categories:

Direct link: The timing reference at the transmitter can be linked *directly* to the timing reference at the receiver. This is illustrated in Fig. 5.28. We can form the link, for example, using a coaxial cable. Because a wired link is impractical for large distances, this form of synchronization is usually reserved for short-range indoor and outdoor radiometers.

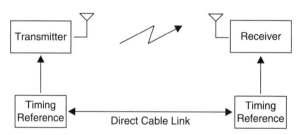

Figure 5.28   One possible transmitter–receiver setup where timing synchronization is maintained via a direct link between timing references at both link ends.

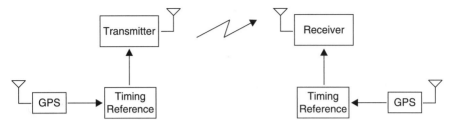

Figure 5.29 Synchronization via indirect link, such as a GPS timing reference.

Global Positioning System (GPS) synchronization: As illustrated in Fig. 5.29, an indirect link can be achieved via the GPS network, which provides a very accurate universal timing signal to all GPS receivers [typically 10^{-12} to 10^{-14} s variance (117)]. This timing signal can, in turn, be used to synchronize the LOs and sample clocks at the transmitter and receiver. This type of synchronization can be used anywhere where there is access to the GPS network, which is justifiably outdoors with a good line-of-sight to the sky. It is important to note that, while GPS synchronization provides an absolute reference, it does not compensate for any phase shift across the channel. Thus, at the receiver, the phase of the transmitter sample clock and local oscillator must be recovered to recover the transmitted signal.

Embedded timing information: As illustrated in Fig. 5.30, extraneous information can be embedded in the transmit signal such that the state of the transmitter timing reference can be inferred at the receiver. Usually, the extraneous information is in the form of a *pilot signal* or *training data*. A pilot signal can take the form of a tone placed just outside of the transmitted bandwidth. This method is used, for example, in analog television broadcast signals. Training data often take the form of a sequence of chips, embedded at regular points in the transmit data, that are known at the receiver. Using this data, it is possible to perform timing and carrier recovery using the received signal alone.

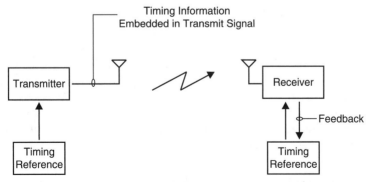

Figure 5.30 Synchronization via active adjustment at the receiver. The receiver infers timing information from the received signal and actively tunes its own timing reference to match its counterpart at the transmitter.

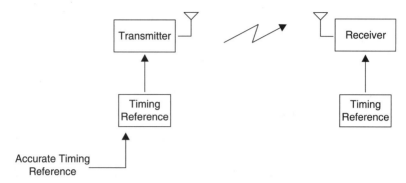

Figure 5.31 Synchronization is provided via accurate timing references at both the transmitter and receiver, whose relative frequency does not change appreciably over the duration of a measurement.

The receiver can use the embedded timing information in one of two ways. It can either *actively* adjust its timing reference via a feedback loop so that some optimization criteria is fulfilled or it can perform time/phase/frequency recovery using a digital signal processing (DSP) method in postprocessing.

Accurate/calibrated timing references: This type of system, illustrated in Fig. 5.31, employs very accurate timing references at the transmitter and receiver. The relative frequency between the two references does not drift appreciably over some useful length of time. In this case, the absolute timing, frequency, and phase at both ends is synchronized for some finite period, and little to no further synchronization is necessary. This form of synchronization is especially popular in radiometers, such as the BYU wideband channel sounder. Accurate timing references include atomic decay clocks, in which the decay of a radioactive isotope such as cesium or rubidium is monitored to measure time. A typical rubidium source, for example, can achieve accuracy of the order 10^{-12} s (117). This type of system is illustrated in Fig. 5.31.

5.7.1 Digital Timing Recovery Methods

This section discusses the case where the state of the sample clock must be recovered in postprocessing. We focus on two timing recovery methods as used in the WMSDR. The first is an adaptive algorithm that involves using a short training sequence. This technique is useful, for example, in packet data transmission. The second method involves transmitting a known ML sequence and cross-correlating that sequence with the received signal. This is useful for timing recovery in correlative channel sounders, where we often use relatively long PN sequences to sound the channel.

5.7.1.1 Adaptive Timing Recovery Long streams of data are often divided into smaller *data packets*. The duration of each packet is usually chosen such that the channel does not change appreciably over its duration. Furthermore,

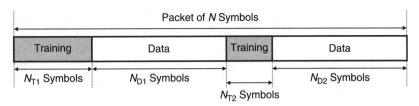

Figure 5.32 An example data packet with embedded training data for timing and carrier recovery.

training symbols are often embedded into each data packet. These training symbols facilitate timing and carrier recovery at the receiver. They can also be used to undo the effects of the channel and thus increase the overall SNR. We refer to symbols not already known at the receiver as *data symbols*.

Figure 5.32 shows an example of how a data packet can be divided into data and training packets. In this example, the data packet consists of $N_D = N_{D1} + N_{D2}$ data symbols and $N_T = N_{T1} + N_{T2}$ training symbols. In general, there can be more than one group of training symbols, inserted at different points within the packet. The size of each group can be fixed or variable. In some cases, the training symbols are encoded throughout the packet. The data symbols represent the information-bearing content of the packet. Thus, given a packet length N, the goal in digital communications is to maximize the ratio N_D/N_T such that we maximize the *data throughput* while still maintaining the ability to recover the timing and carrier information.

Figure 5.33 shows a typical sampled waveform at the receiver. In this example, the receiver oversamples the transmit waveform such that it has at least four samples per transmitted symbol. This is an example of *four times oversampling*. The implication is that there are more samples than necessary; only one sample per pulse is required.

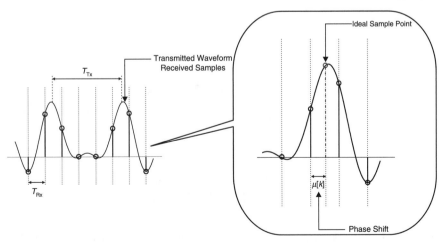

Figure 5.33 An example of a transmitted waveform and received samples where the receiver oversamples the transmit waveform by four times.

However, from DSP theory (61), oversampling increases dynamic range by decreasing the noise floor. Therefore, the extra samples can be used to increase the receive SNR.

To maximize the SNR at the receiver, a point at the ideal sample instant can be interpolated. In Fig. 5.33, the ideal sample point is located at exactly μ_k away from a reference receiver sample. Given some method of determining $\mu[k]$, an *interpolator* can be used to interpolate a point at the ideal sample instant. Thus, $\mu[k]$ describes the (possibly time-varying) phase difference between the transmitter and receiver clocks.

5.7.1.2 The Gardner Timing Algorithm

The *Gardner timing algorithm* is a metric that quantifies how far the receiver sample instant is off from the ideal point. The algorithm is used in many digital electronics such as phone modems and cell phones. This metric, in turn, can be used to compute $\mu[k]$.

The Gardner timing algorithm requires at least two samples per transmitted pulse, or at least two times oversampling. In this case, the kth Gardner error metric $e_{\text{Gardner}}[k]$ is computed as

$$e_{\text{Gardner}}[k] = [r[k+2] - r[k]]r[k+1], \tag{5.49}$$

where $r[k]$ is the kth receive sample. Figure 5.34 illustrates three possible cases at the receiver. Each subplot shows the transmit waveform and receive samples.

Figure 5.34a demonstrates the case where the receiver clock *leads* the transmitter clock. In Fig. 5.34b, the receive clock *lags* the transmitter clock. In Fig. 5.34b, both the transmit-clock and receive-clock are perfectly synchronized. For the first case, $e_{\text{Gardner}}[1]$ is computed as

$$\begin{aligned} e_{\text{Gardner}}[1] &= [r[3] - r[1]]r[2] \\ &= [-0.5 - 0.5]0.5 \\ &= -0.5. \end{aligned} \tag{5.50}$$

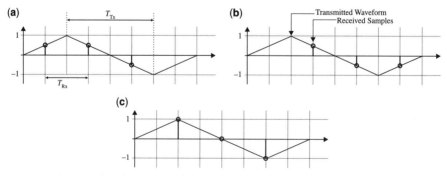

Figure 5.34 Example sampled waveforms to illustrate the Gardner timing algorithm error metric. (a) Receiver leads transmitter. (b) Receiver lags transmitter. (c) Receiver samples at ideal time instants.

Therefore, when $e_{\text{Gardner}}[k] < 0$, the receiver clock leads the transmitter clock. Similarly, for the second case,

$$
\begin{aligned}
e_{\text{Gardner}}[1] &= [-0.5 - 0.5](-0.5) \\
&= 0.5.
\end{aligned}
\tag{5.51}
$$

so that when $e_{\text{Gardner}}[k] > 0$, the receiver clock lags the transmitter clock. Finally, when the clocks are synchronous, we have

$$
\begin{aligned}
e_{\text{Gardner}}[1] &= [-1 - 1](0) \\
&= 0.
\end{aligned}
\tag{5.52}
$$

The goal therefore is to adjust the sample timing until $e_{\text{Gardner}}[k] = 0$. Note that the algorithm works best when the transmitted signal alternates its amplitude at $1/T_{\text{Tx}}$. This guarantees that any timing recover algorithm that uses $e_{\text{Gardner}}[k]$ to adjust the receiver sample period will converge to T_{Tx} in the shortest time possible.

5.7.1.3 An Adaptive Timing Recovery Algorithm
The following describes an adaptive timing recovery algorithm, which is a variation of the one described in Ref. 118, adapted for use with the WMSDR. The algorithm assumes that the data are divided into packets. Furthermore, each packet has at least one training sequence. In addition to timing recovery, the algorithm also decimates the received signal so that, at its output, there is one sample per transmit pulse.

Because it uses the Gardner timing algorithm as its feedback metric, the algorithm assumes that the receiver oversamples the transmitter by at least two times. The oversampling rate does not necessarily have to be an integer. The algorithm works by adjusting a phase and *control word* based on the Gardner error. The algorithm then updates an index and phase variable based on the control word. The control word keeps track of the absolute timing difference between the transmitter and receiver.

Consider the simplified example shown in Fig. 5.35. The transmitted signal is represented as the solid curve. The receiver oversamples the transmitted signal by four times.

Let $\hat{s}_{\text{Tx}}[k, j]$ denote the kth estimate of the jth transmit sample period, where the index j is relative to the current receiver sample $r[k]$. $\hat{s}_{\text{Tx}}[k, j]$ is most often not an

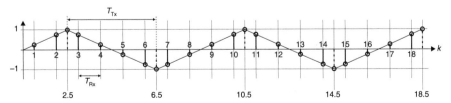

Figure 5.35 Example waveform showing transmitted waveform and received samples where the receiver oversamples the transmit waveform by four times.

integer; the transmitter sample time can lie between received samples. For example, referring to Fig. 5.35, the first sample period lies midway between receive samples 2 and 3.

$\hat{s}_{Tx}[k, j]$ is computed using the iterative equation

$$\hat{s}_{Tx}[k + 1, j] = m[k] + \mu[k] + (j - 1)\frac{w[k]}{R_{out,over}}, \tag{5.53}$$

where $m[k]$ is the index of the current receiver sample, $\mu[k]$ describes the phase difference between the transmitter and receiver clocks, and $w[k]$ is the *control word*. As the algorithm progresses, $w[k]$ eventually converges to the oversample rate. The constant $R_{out,over}$ determines the oversampling rate at the output of the algorithm. That is, the oversampling rate can be changed such that $1 \leq R_{out,over} \leq R_{Rx,over}$, where the oversampling rate of the receiver is $R_{Rx,over} = T_{Tx}/T_{Rx}$. The index variable $m[k]$ and the phase variable $\mu[k]$ are updated with each iteration of the algorithm. Note that $\hat{s}_{Tx}[k, j]$ is a function of two indices, k and j. The index k refers to the current receiver sample $r[k]$. Using (5.53), a point closest to $r[k]$ or at multiples of T_{Tx} away from $r[k]$ can be interpolated. This is done by varying the index j. For example, setting $j = 0$, $\hat{s}_{Tx}[k, 0]$ is a point $m[k] + \mu[k]$ away in time from the current receiver sample $r[k]$. By setting $j = 0$, $\hat{s}_{Tx}[k, 1]$ is a point $w[k]/R_{out,over}$ away from $\hat{s}_{Tx}[k, 0]$ in time. As the algorithm progresses, the ratio $w[k]/R_{out,over}$ converges to T_{Tx}.

The algorithm can be described by the following steps:

1. Initialize $m[k]$, $\mu[k]$, and $w[k]$ to the best-guess estimate of the integer and fraction location of the first transmitted sample and receiver oversample rate, respectively.
2. Compute $\hat{s}_{Tx}[k, j]$.
3. Interpolate the three transmitter samples at $\hat{s}_{Tx}[k, j]$ for $j = 1 \ldots 3$.
4. Using these three interpolated points, compute the timing error estimate $e_{Gardner}[k]$.
5. Filter the error estimate according to

$$e_{filter}[k] = k_p(e_{Gardner}[k] - e_{Gardner}[k - 1]) + k_i \cdot e_{Gardner}[k - 1]$$
$$+ e_{filter}[k - 1], \tag{5.54}$$

where $0 < k_i, k_p \leq 1$ are arbitrary constants that control the rate of convergence and overshoot of the algorithm.

6. Update the control word $w[k]$ according to

$$w[k] = w[k - 1] + e_{filter}[k]. \tag{5.55}$$

7. Update $m[k]$ according to

$$m[k] = m[k - 1] + \text{floor}(\mu[k - 1] + w[k]), \tag{5.56}$$

where floor(x) yields the integer part of x.

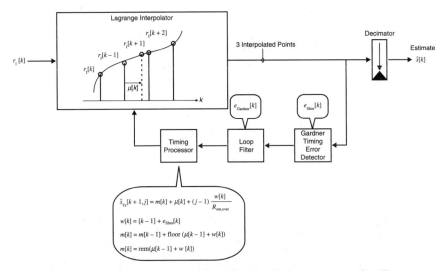

Figure 5.36 Block diagram of the adaptive timing recovery algorithm.

8. Update $\mu[k]$ according to

$$\mu[k] = \text{rem}(\mu[k-1] + w[k]), \tag{5.57}$$

where $\text{rem}(x)$ yields the decimal part of x. A block diagram of the algorithm is shown in Fig. 5.36.

5.7.1.3.1 Results Using WMSDR Data The following demonstrates a few examples, where the above adaptive timing algorithm is used with real-life WMSDR data. The hardware setup is shown in Fig. 5.37. At the transmitter, each data packet consisted of $N_T = 500$ training and $N_D = 1000$ data chips. This information was encoded using one of five different signal constellations: BPSK (binary phase-shift keying), QPSK (quadrature PSK), 8-PSK, 16-QAM (quadrature amplitude modulation), and 36-QAM. The chip rate was fixed at 4 MChips/s. The receiver sampling rate was set to 10 MHz, which meant that the transmitted signal was oversampled by 2.5 times.

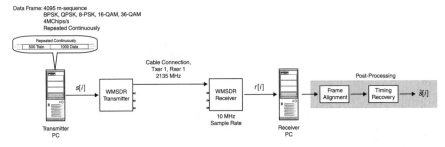

Figure 5.37 Adaptive timing algorithm test setup using the WMSDR.

The transmitted signal constellation was recovered from the received samples in postprocessing using the above adaptive algorithm. In addition, we also recovered the phase of the constellation using the algorithm outlined in Section 5.7.2.

Figure 5.38 illustrates an example receive packet, where the random data have been encoded using 16-QAM. The dashed line indicates the received signal, and the stem plot indicates the interpolated transmit sample points. Figure 5.38d shows a portion of the data sequence. Figure 5.39 shows the received signal after timing recovery for the BPSK, QPSK, 8-PSK, 16-QAM, and 36-QAM constellations.

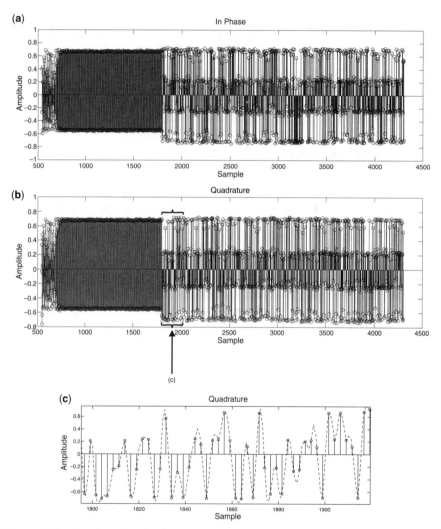

Figure 5.38 Showing the in-phase and quadrature received signals and time-recovered samples where the samples were recovered using the adaptive timing algorithm. The data were transmitted using 16-QAM.

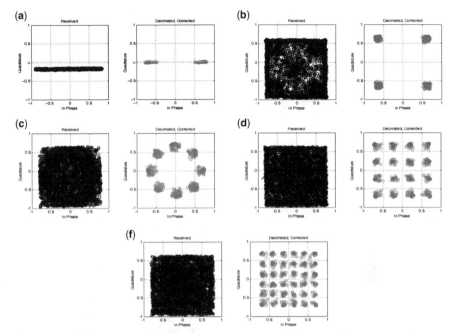

Figure 5.39 Constellations using the adaptive timing algorithm with WMSDR data and different constellations: (a) BPSK, (b) QPSK, (c) 8-PSK, (d) 16-QAM, (e) 36-QAM.

The adaptive timing algorithm outlined here suffers from a number of drawbacks. First, the algorithm is complex and relies on user-selected constants such as k_i and k_p, which need to be adjusted for every new channel. Second, and perhaps most seriously, the algorithm will not converge if the data stream undergoes time dispersion. This makes it impractical for use in wideband channel sounding. The convergence time of the algorithm is very sensitive to received SNR; the lower the receive SNR, the longer the algorithm takes to converge.

5.7.1.4 A Timing Recovery Method Using Cross-correlation

Unique to the problem of correlative channel sounding is the fact that the transmitted sequences are known at the receiver. In this way, the entire transmitted sequence can be thought of as a training sequence. Thus, one of the simplest ways to recover the symbol timing at the receiver is to cross-correlate the known transmitter sequence with the received signal.

In Ref. 119, Paier et al. describe a timing recovery algorithm designed to work in time-dispersive channels. The algorithm is robust and simple to implement. Furthermore, this algorithm can be used in a MIMO channel, in which there are M_{Tx} superimposed signals. The algorithm assumes that the received signal is an oversampled version of the transmitted sequence. The higher the oversample rate at the receiver, the better the timing estimate.

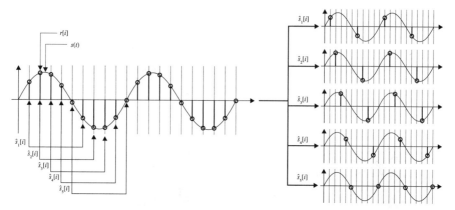

Figure 5.40 An example of generating several interpolated sequences from one oversampled signal. The interpolated sequence all have $f_s = 1/T_{Tx}$.

Figure 5.40 illustrates the case where the transmitter sends the sequence $s[i] = [1, -1, 1, -1]$. Here, the receiver records five samples per transmitted pulse. From this, five different *interpolated sequences* $\hat{s}_j[i]$ can be created. Each $\hat{s}_j[i]$ is formed by taking samples T_{Tx} apart. Each waveform is delayed by T_{Rx} from the last. Each interpolated sequence can be cross-correlated with the known transmitted sequence, and the interpolated sequence associated with the highest correlation represents the one closest to the actual transmitter symbol timing.

We compute the *error metric*, m_j, by cross-correlating each of the $\hat{s}_j[i]$ with the original sequence $s[i]$, viz.,

$$m_j = \sum_{i=1}^{L_{Tx}} |\hat{s}_j[i]s[i]|^2, \tag{5.58}$$

where L_{Tx} is the length of $s[i]$. After computing m_j for all j, we pick the $\hat{s}_j[i]$ associated with the *largest m_j*. The result is that we choose the closest approximation to $s[i]$.

5.7.1.4.1 Example Using WMSDR Data   In the following, we verify the performance of the above timing recovery algorithm using WMSDR data. The setup is illustrated in Fig. 5.41. One antenna port at the transmitter is connected directly to a receiver antenna port. The transmitter is programmed to transmit a 4096 chip ML-sequence at 4 MChips/s using a BPSK constellation. The receiver recorded complex baseband samples at 8 MHz, which meant it oversampled the transmitter by two times. The above cross-correlation timing recovery algorithm was implemented in postprocessing. Note that in the following plots, we did not recover the phase of the carrier, only the symbol timing.

First off, a digital filter is used to *upsample* the received waveform by four times. This brings the total oversample rate to eight times. Figure 5.42 illustrates a portion of the upsampled received in-phase and quadrature waveforms $r_{I,up}[i]$ and $r_{Q,up}[i]$. The quadrature waveform is 180° out from the in-phase waveform and noticeably smaller in amplitude. This is due to the LO phase offset.

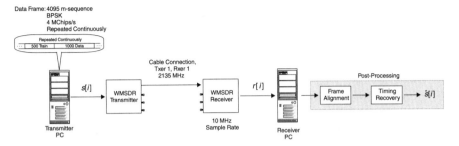

Figure 5.41 Test setup for timing recovery using WMSDR data.

The transmitted sequence $s[i]$ is known at the receiver. The interpolated sequences $\hat{s}_{I,j}[i]$ and $\hat{s}_{Q,j}[i]$ are derived from $r_{I,up}[i]$ and $r_{Q,up}[i]$, respectively. Figure 5.43 illustrates $\hat{s}_{I,j}[i]$ plotted against $r_I[i]$, for all $j \in \{1, \ldots, 8\}$. The metric m_j is plotted in Fig. 5.44. Note that $\hat{s}_{I,5}[i]$ is the interpolated sequence that yields the largest m_j and is thus the closest to the original sequence $s_I[i]$.

Figure 5.45 shows the resulting received signal constellation, where each $\hat{s}_{I,5}[i]$ and $\hat{s}_{Q,5}[i]$ is plotted as a single point on the IQ-plane. The received constellation shown is not perfect. The constellation points are "smeared," both radially and in the azimuth. The smearing in azimuth is due to a frequency difference between the receive LO and the transmit LO. Also, the received constellation is rotated with respect to the transmitted constellation because there is a phase difference between the two LOs. The radial smearing is due to a number of factors. The most notable of them are the

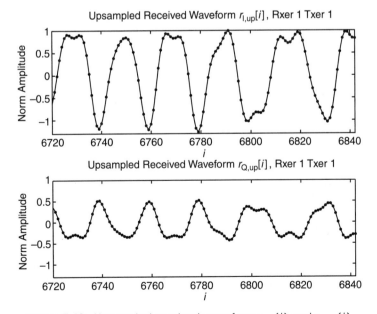

Figure 5.42 Upsampled received waveform $r_{I,up}(i)$ and $r_{Q,up}(i)$.

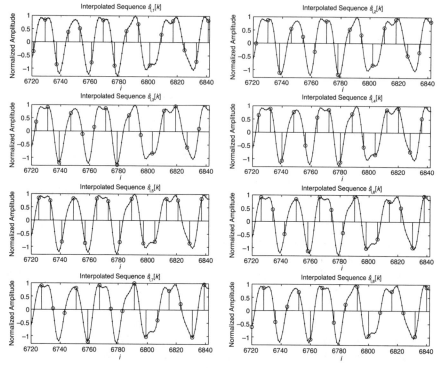

Figure 5.43 All in-phase interpolated sequences $\hat{s}_{1,j}(i)$ for the above example.

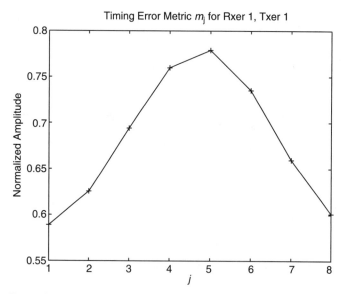

Figure 5.44 Cross-correlation error metric m_j for above example.

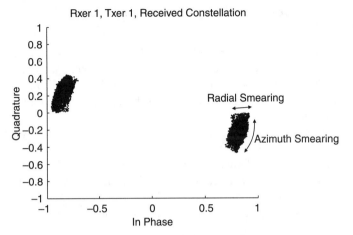

Figure 5.45 Received signal constellation $\hat{s}_{I,5}(i)$ versus $\hat{s}_{Q,5}(i)$ before phase recovery.

timing discrepancy, presence of noise in both the transmitter and receiver electronics, and a mismatch between the transmit-baseband and receive-baseband filters.

The cross-correlation timing recovery algorithm suffers from a few drawbacks. First, oversampling the transmit data requires either increased hardware complexity at the receiver, increased computing complexity in postprocessing, or both. For example, in the above results, the WMSDR receiver worked at an oversample rate of two times the transmitter chip rate. Increasing this rate would reduce the timing error of the estimated sequence. However, this also increases the amount of recorded data. Both the sample rate and data capacity/throughput are limited by the receiver hardware.

Second, for the algorithm to work, the frequency of the receiver sample clock cannot drift appreciably over the duration of the training sequence. Recall that the time between samples in $\hat{s}_j[k]$ is the approximate transmit sample period $\hat{T}_{\text{Tx}}$, where

$$\hat{T}_{\text{Tx}} = R_{\text{Rx,upsample}} \cdot T_{\text{Rx}}. \tag{5.59}$$

The total oversample rate at the receiver $R_{\text{Rx,upsample}}$ is the oversample rate of the receiver $R_{\text{Rx,over}}$ multiplied by the upsample rate in postprocessing R_{upsample}. More specifically,

$$R_{\text{Rx,upsample}} = R_{\text{Rx,over}} \cdot R_{\text{upsample}}. \tag{5.60}$$

The above algorithm assumes that $\hat{T}_{\text{Tx}} \simeq T_{\text{Tx}}$. If the receive-clock frequency drifts relative to the transmit clock, then T_{Rx} is a function of time, and the above relations do not hold. Also, in order for the above relations to work, T_{Tx} must be an integer multiple of T_{Rx}.

5.7.2   Phase Recovery Using a Decision-Directed Feedback Loop

In the case of *maximum a-posterior* (MAP) *probability* receivers, it is often required that the receive constellation points are centered within certain boundaries on the IQ-plane (106). This implies that, in addition to recovering the sample timing, the *carrier phase* must be recovered. Figure 5.46 illustrates the example where we have a rotated QPSK constellation point. The following section outlines a simple algorithm that tracks the carrier phase on a chip-by-chip basis. It assumes that the symbol timing has already been recovered.

A block diagram of the phase recovery algorithm is shown in Fig. 5.47. To explain, first, it is assumed that the ideal constellation points are known at the receiver. We denote each constellation point s_c for $c \in \{1, \ldots, C\}$, where C is the total number of constellation points. Each s_c can be viewed as a complex number,

$$s_c = s_{I,c} + js_{Q,c}, \tag{5.61}$$

where $s_{I,c}$, $s_{Q,c}$ are the in-phase and quadrature amplitudes of the cth constellation point, respectively. Table 5.1 lists s_c for the QPSK constellation. The received sample $r[i]$ also lies on the complex IQ-plane and can be expressed as

$$
\begin{aligned}
r[i] &= r_I[i] + j \cdot r_Q[i] \\
&= |r[i]| e^{j\theta[i]},
\end{aligned}
\tag{5.62}
$$

where $\theta[i]$ is the phase offset of the ith receive sample $r[i]$ with respect to $s[i]$.

In general, the phase offset can vary over time. For each receive sample, the *closest* constellation point is found, and the error metric $e_\phi[i]$ between the receive point and

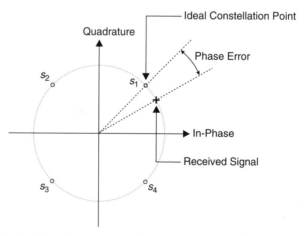

Figure 5.46   Illustration of phase error between a received sample and the ideal constellation point.

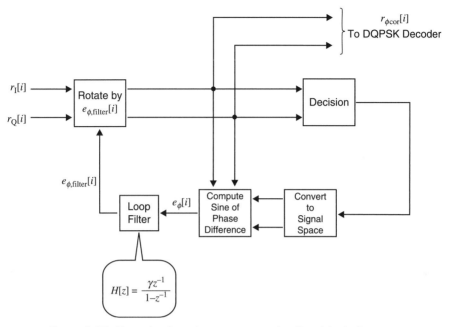

Figure 5.47 The adaptive phase-recovery algorithm block diagram.

the closest constellation point is computed. This metric is computed as

$$e_\phi[i] = s_{I,k} \cdot r_Q[i] - s_{Q,k} \cdot r_I[i], \tag{5.63}$$

where $s_k = s_{I,k} + j s_{Q,k}$ denotes the closest ideal constellation point to $r[i]$. The error is modified by a loop filter, whose transfer function $H_{\mathrm{loop}}[z]$ is given by

$$H_{\mathrm{loop}}[z] = \frac{\gamma \cdot z^{-1}}{1 - z^{-1}}. \tag{5.64}$$

The user-selected constant $0 < \gamma < 1$ determines how quickly the algorithm converges and the noise in the estimate. From this result, the iterative equation for the

Table 5.1 QPSK constellation points represented as complex numbers

c	s_c (Polar)	s_c (Cartesian)
1	$\sqrt{2}e^{j\frac{\pi}{2}}$	$1 + j$
2	$\sqrt{2}e^{j\frac{3\pi}{2}}$	$-1 + j$
3	$\sqrt{2}e^{j\frac{5\pi}{2}}$	$-1 - j$
4	$\sqrt{2}e^{j\frac{7\pi}{2}}$	$1 - j$

filtered error is given by

$$e_{\phi,\text{filter}}[i] = \gamma \cdot e_\phi[i-1] + e_{\phi,\text{filter}}[i-1]. \tag{5.65}$$

The filtered error is then used to correct the phase of $r[i]$, that is,

$$r_{\phi\,\text{cor}}[i] = |r[i]|e^{j(\theta[i]-e_{\phi,\text{filter}}[i-1])}, \tag{5.66}$$

where $r_{\phi\text{cor}}[i]$ is $r[i]$ after phase correction. The error estimate is updated for every new $r[i]$.

5.7.2.1 Phase Ambiguity   One of the biggest disadvantages of phase recovery using a decision-directed feedback loop over other phase estimation algorithms is the resulting *phase ambiguity*. Consider the following example in which a single QPSK chip is transmitted. Let the transmitted chip be

$$s[i] = \sqrt{2}e^{j\frac{\pi}{4}}. \tag{5.67}$$

At the receiver, the path length is such that the phase difference is exactly $\pi/2$, so that the received sample is

$$r[i] = \sqrt{2}e^{j\frac{3\pi}{4}}. \tag{5.68}$$

In this case, the above algorithm determines the nearest ideal constellation point to be $s_2 = \sqrt{e}^{j\frac{3\pi}{2}}$, and the error is computed as $e_\phi[i] = 0$. Assuming that the phase offset $\theta[i]$ does not change appreciably from time instant i to $i+1$, the next phase estimate will also be off by $\pi/2$. The result is that, at the output of the algorithm, the entire constellation will be rotated by $\pi/2$ with respect to the transmitted constellation. In the literature, this is often referred to as a $\pi/2$ phase ambiguity. The size of the ambiguity in the azimuth depends on the order of the constellation. For example, for a BPSK constellation, there is a π ambiguity, for an 8-PSK constellation, a $\pi/4$ ambiguity, and so forth. We can overcome this ambiguity either by using a short training sequence or by using *differential decoding* methods, such as differential-QPSK (109). Another disadvantage is that the algorithm takes some time to converge. Thus, a training sequence has the advantage of resolving the ambiguity, in addition to allowing the algorithm some time to converge.

5.7.2.2 A Simple Example   The following demonstrates the utility of the phase recovery algorithm using a simple MATLAB example. At the transmitter, a random QPSK constellation is generated. The receiver records a noisy and phase-shifted version of the transmitted signal, as shown in Fig. 5.48. For the above plots, a 1000 chip transmit sequence was used, the receive SNR was chosen to be 20 dB, and the loop filter constant was $\gamma = 0.2$. Also, the received constellation phase offset was constant; that is, we let $\theta[i] = 30°$. Figure 5.49 shows the filtered error

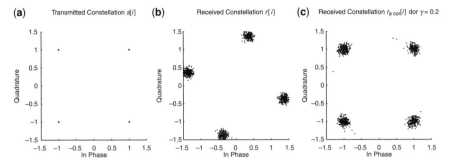

Figure 5.48 An example of a phase-recovered signal constellation. (a) The transmitted constellation. (b) The received constellation. (c) The received constellation after phase recovery.

$e_{\phi,\text{filter}}[i]$ for the entire sequence. From this figure, it is evident that the error converges rapidly to its steady state.

The noise in the steady-state solution is due to both noise at the receiver and errors in the estimate. The variance in the steady state can be reduced by decreasing γ. For example, Fig. 5.50a plots $e_{\phi,\text{filter}}[i]$ for the same conditions, with $\gamma = 0.01$. The steady-state solution is less noisy, but it takes much longer for the algorithm to converge. This tends to smear the receive constellation points, as shown in Fig. 5.50b.

5.7.2.3 Results Using WMSDR Data The above algorithm was applied to the WMSDR data. The example, presented herein, used a BPSK-encoded ML sequence. The time-corrected constellation, shown in Fig. 5.51a, was rotated and

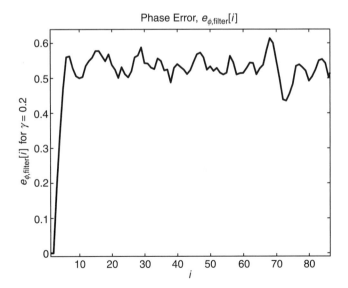

Figure 5.49 $e_{\phi,\text{filter}}(i)$ for the above example, for which $\gamma = 0.2$.

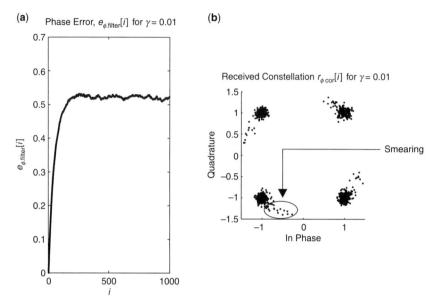

Figure 5.50 Illustration of the receive constellation after phase recovery, with $\gamma = 0.1$. (a) Filtered error $e_{\phi,\text{filter}}(i)$. (b) Received constellation after phase recovery. The increased number of smeared constellation points is due to the increase convergence time of the algorithm.

smeared in the azimuth due to a phase and frequency offset between the transmit-LO and receive-LO.

Figure 5.51 shows the received constellation before and after phase recovery. Note that there was only slight azimuth smearing, caused by the finite convergence time. Figure 5.52 is a plot of $e_{\phi,\text{filter}}[i]$. In this case, $e_{\phi,\text{filter}}[i]$ exhibits a larger variance

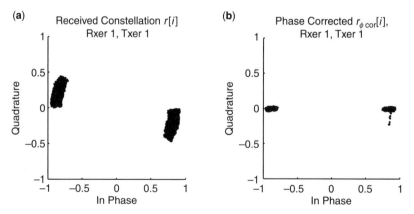

Figure 5.51 Received signal constellation (a) before phase correction and (b) after phase correction, using WMSDR data.

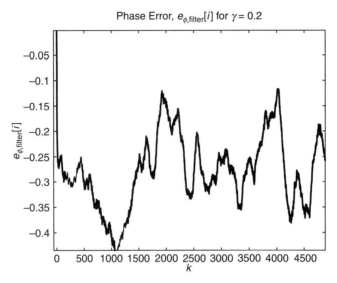

Figure 5.52 Filtered error $e_{\phi,\text{filter}}(i)$ for the above example.

than in the MATLAB example above. This is mainly due to noise sources in the measurement apparatus, as well as a noisy sample time estimate, evidenced by the radial "smearing" of the constellation points.

5.8 SUMMARY AND DISCUSSION

A channel sounder is used to obtain unbiased estimates of the channel impulse response. The wideband MIMO software defined radio (WMSDR) is an example of a 4×4 wideband MIMO channel sounder capable of simultaneously measuring the response of all parallel channels. Although the WMSDR is capable of many tasks, in this text we focus primarily on estimating the impulse response of outdoor channels.

There are many channel sounding methods. The type of channel being measured and the capabilities of the channel sounder dictate the method used. Periodic pulse sounding is perhaps the most general form of channel sounding but is difficult to implement. Narrowband SISO and MIMO channel sounding use pilot tones to estimate the response in a small portion of the band. Correlative channel sounding uses PN sequences, and a correlator, to estimate the wideband channel.

Digital matched filtering is a correlative sounding technique. It computes calibrated estimates of the time-invariant channel using DSP techniques. Calibration is used to reduce the frequency distortion introduced by the measurement apparatus. The digital matched filtering technique can be extended to include the MIMO case. This technique is used to estimate the channel H-matrix from WMSDR data in Chapter 6.

When using coherent detection schemes, the chip timing must be recovered at the receiver. We presented two timing recovery algorithms. The first algorithm works iteratively and requires an embedded training sequence. This algorithm is not suited to wideband MIMO channel sounding for the following reasons:

- The algorithm cannot distinguish between superimposed signals.
- It cannot be used in channels that exhibit significant time-dispersion.

We therefore presented a second timing recovery algorithm that used cross-correlation to recover the chip timing from the received sequence. This algorithm is simple, robust, and well suited for use in the wideband MIMO channel.

When using coherent detection schemes, carrier recovery must also be performed. In the past section, we presented a simple phase recovery algorithm. The algorithm suffered from phase ambiguity, which could be resolved by using a differential encoding scheme or a short training sequence.

5.9 NOTES AND REFERENCES

There are a number of examples of MIMO channel sounders, both narrowband and wideband, in the literature. One of the first MIMO channel sounders, an 8×12 narrowband system capable of transmitting and receiving digital data, was developed by Golden et al. at Bell Labs (120, 121). To maintain synchronicity, the receiver and transmitter were linked together by coaxial cable. This restricted the system range. The system was mainly used indoors to test the *vertical-Bell Labs layered space–time* (V-BLAST) and, later, the Turbo-BLAST (21) communication algorithms. A summary of the BLAST project can be found online (122).

Wolniansky et al. were the first to build a 16×16 narrowband test set dubbed the *vector-space narrowband radiometer* (VSNR). It was subsequently used in several measurement campaigns in midtown Manhattan (26). As part of his graduate work and with help from Bell Labs, Aron (123) built a replica of the VSNR and used it in several measurement campaigns. His thesis contains a detailed description of the VSNR. The VSNR works by simultaneously transmitting 16 pilot tones using a series of *direct digital synthesizers* (DDSs) but is not able to transmit data.

Perhaps the most popular architectures for MIMO sounders is the switched-array type. This is mostly due to the reduced hardware complexity, and increased bandwidth, versus true MIMO sounders. A few switched-array sounders found in the literature include the MEDAV sounder (115), the Elektrobit PropSound CS (124), and BYU wideband channel sounder (116). The first two sounders are commercially available, whereas the last was an experimental apparatus built at BYU. The Wireless Communications Research Lab at BYU also produced a 16×16 narrowband sounder (125), as well as a wideband SISO sounder designed to measure the AoA of received signals (94).

6

EXPERIMENTAL VERIFICATIONS

In this chapter, we present an experimental analysis of two correlative channel models; namely, the structured and Kronecker models. There is only a handful of correlative wideband MIMO channel models in the literature. As previously pointed out, the wideband Kronecker model is by far the most popular, mainly due to the separability assumption and therefore the simplicity of the model. We compare the performance of the structured versus Kronecker models using real-life channel data, gathered from the wideband MIMO software defined radio (WMSDR) and Brigham Young University (BYU) wideband channel sounder.

The chapter begins with a discussion of a few metrics commonly used to validate the performance of MIMO channel models. It then describes the experimental setup for the WMSDR and BYU data sets. Both sets of data were collected in very different environments, using two apparatuses whose specifications differ greatly. The WMSDR data were recorded in a fixed point-to-point outdoor environment. For most locations, the receiver had direct line-of-sight to the transmitter. In contrast, the BYU data were recorded using a much higher bandwidth than that of the WMSDR. The BYU data were recorded in a fixed point-to-point indoor environment, where no line-of-sight existed between the transmitter and receiver. We show the robustness of the structured model using both of these data sets.

Using the MIMO azimuth power spectrum (APS), we qualitatively show how the structured model better approximates the spatial structure of the channel. One of the structured model assumptions is that the coupling between scatterers at different delays is significant. We discuss this issue and plot the wideband correlation for two typical channels.

Multiple-Input, Multiple-Output Channel Models: Theory and Practice. By Nelson Costa and Simon Haykin
Copyright © 2010 John Wiley & Sons, Inc.

6.1 VALIDATION METRICS

This section covers a few metrics used in the literature to verify the performance of MIMO channel models. Each metric is unique in that it tests a different aspect of the model. This section provides an interpretation of each metric and its intended use. The channel capacity metric is used in the last half of the chapter to compare the performance of the structured and Kronecker models.

6.1.1 Channel Capacity

The MIMO capacity is by far the most popular metric used in the literature, perhaps because it is arguably the most important. MIMO capacity is not a good indication of the spatial structure of the channel (80). However, the capacity of a given MIMO channel is often a critical consideration for higher-layer operation; for example, to estimate throughput at the packet layer and overall end–end performance in a simulated channel. Also, the growing acceptance of MIMO technology is based largely on the promise of increased spectral efficiency. Thus, the accuracy with which a model characterizes the capacity of real-life channels, in large part, determines its usefulness in practice.

The *instantaneous capacity*, or the capacity at any given time instant, for a *narrowband* MIMO channel is governed by *Foschini's log-det formula* (19),

$$C_H = \log_2\left[\det\left(\mathbf{I} + \frac{\rho}{M_{\text{Tx}}}\mathbf{H}\mathbf{H}^H\right)\right], \tag{6.1}$$

where $\mathbf{H} \in \mathbb{C}^{M_{\text{Rx}} \times M_{\text{Tx}}}$ is the narrowband H-matrix, $\det(\cdot)$ is the matrix determinant, and C_H is the *instantaneous narrowband capacity*, in bits $\text{s}^{-1}\,\text{Hz}^{-1}$ or bps/Hz. The system signal-to-noise Ratio (SNR), ρ, is chosen such that the measurement SNR is guaranteed to be greater than the system SNR for all measurements (26). Equation (6.1) assumes perfect *channel state information* (CSI) at the receiver, but no CSI at the transmitter. Without CSI at the transmitter, the best transmission scheme is to allot power equally across all transmit antennas (25).

Assuming that the channel is wide-sense stationary, the *ergodic narrowband capacity* $\overline{C}_H$ can be computed as the expected value of the instantaneous capacity over time; that is,

$$\overline{C}_H = \mathrm{E}\{C_H\}. \tag{6.2}$$

To evaluate wideband models properly, a formula for the capacity of a wideband MIMO channel is required. In the literature, the wideband MIMO channel is most often viewed as a collection of D narrowband MIMO channels, where D is the maximum delay. Stated differently, the wideband H-matrix $\mathbf{H}[d]$ can be viewed as a narrowband H-matrix for each $d \in \{1, \ldots, D\}$. In this case, Foschini's formula can be extended to include the wideband case by simply summing the capacity contribution of all D narrowband MIMO channels. Thus, in the literature, the instantaneous capacity for the wideband MIMO channel is most often computed using (46):

$$C_X = \frac{1}{W} \sum_{f=1}^{F} \log_2 \left[\det \left(\mathbf{I} + \frac{\rho}{M_{\text{Tx}}} \widetilde{\mathbf{H}}[f] \widetilde{\mathbf{H}}^{H}[f] \right) \right] \Delta f, \qquad (6.3)$$

where C_X is the instantaneous wideband MIMO capacity in bps, W is the channel bandwidth in Hz, F is the number of frequency bins, ρ is the system SNR, $\Delta f = W/F$, $\widetilde{\mathbf{H}}[f]$ is the normalized wideband $\mathbf{H}[f]$, and $\mathbf{H}[f] \underset{\text{FFT}}{\leftrightharpoons} \mathbf{H}[d]$. $\mathbf{H}[f]$ is normalized via the constant n_X,

$$n_X = \sqrt{\frac{1}{M_{\text{Rx}} M_{\text{Tx}} F} \sum_{i=1}^{M_{\text{Rx}}} \sum_{j=1}^{M_{\text{Tx}}} \sum_{k=1}^{F} |h_{ijk}|^2}, \qquad (6.4)$$

where h_{ijk} is the (i, j, k)th element of $\mathbf{H}[f]$. The square of the normalization constant n_X^2 can be viewed as the average complex gain, averaged over all elements in $\widetilde{\mathbf{H}}_X[f]$, for all $f \in \{1, \ldots, F\}$. The $\mathbf{H}[f]$ for each snapshot of the channel is normalized independently. This means that the instantaneous SNR remains constant over the duration of the measurement or simulation. In a real-life scenario, this would be the equivalent of perfect power control at the transmitter, where the transmitter regulates its instantaneous power to maintain a constant SNR at the receiver.

In the above equations, ρ requires practical consideration; the system SNR is a function of the power restriction at the transmitter and the noise variance. That is, let $\mathbf{x}$ represent the $M_{\text{Tx}} \times 1$ transmit vector. The transmit power is constrained such that $E\{\mathbf{x}\mathbf{x}^{H}\} = P$. Also, it is often assumed that the $M_{\text{Rx}} \times 1$ noise vector $\mathbf{n}$ has zero-mean complex Gaussian distributed elements, each with variance $E\{\mathbf{n}\mathbf{n}^{H}\} = \sigma^2$. The system SNR is thus defined as $\rho = P/\sigma^2$. In this way, ρ is sometimes referred to as the *transmit SNR*, as it is the SNR seen at the transmitter, before the transmit signal is affected by the channel. In general, as ρ increases, so does C_X. Upon further inspection of the literature, a popular choice seems to be $\rho = 20$ dB; for example, see Refs. 46 and 126.

6.1.2 The Diversity and Correlation Metrics

The *diversity metric* is a measure of the number of independently faded paths in a MIMO channel. In was first presented in Ref. 127 as a method for quantifying the diversity and correlation in Rayleigh fading MIMO environments. It was subsequently used in Ref. 80 to validate the random cluster model (RCM) using data from several different measurement campaigns and is included in the COST 273 framework (91). The diversity metric can be used to rank channels according to their diversity level. It was intended as a simple metric that would allow the transmitter or receiver to make an informed choice regarding the use of a diversity or spatial multiplexing scheme.

Given the narrowband correlation matrix $\mathbf{R}_H \in \mathbb{C}^{M_{Rx} \times M_{Tx}}$, the diversity measure $\Psi(\mathbf{R}_H)$ is defined as (127)

$$\Psi(\mathbf{R}_H) = \left(\frac{\text{tr}(\mathbf{R}_H)}{\|\mathbf{R}_H\|_F} \right)^2, \tag{6.5}$$

where $\text{tr}(\cdot)$ is the trace operator, and $\|\cdot\|_F$ denotes the Frobenius norm.

The diversity metric has several interesting properties. If the channels were ranked according to their diversity metric, channels with similar diversity metrics offer similar ergodic capacities. When the channel is perfectly *uncorrelated*, $\mathbf{R}_H = \mathbf{I}_{M_{Rx} \times M_{Tx}}$, and $\Psi(\mathbf{R}_H) = M_{Rx}M_{Tx}$. When the channel is perfectly correlated, for example, all entries of $\mathbf{R}_H = \rho$, where ρ is some arbitrary constant, $\Psi(\mathbf{R}_H) = 1$. Thus

$$1 \leq \Psi(\mathbf{R}_H) \leq M_{Rx}M_{Tx}. \tag{6.6}$$

In addition, $\Psi(\mathbf{R}_H)$ is closely related to the eigenvalue profile of $\mathbf{R}_H$. Let λ_i be the eigenvalues of $\mathbf{R}_H$, where $i \in \{1, \ldots, M_{Rx}M_{Tx}\}$. From Ref. 128, we have,

$$\text{tr}(\mathbf{R}_H) = \sum_{i=1}^{M_{Rx}M_{Tx}} \lambda_i. \tag{6.7}$$

Using (6.7), it is easy to show that

$$\|\mathbf{R}_H\|_F^2 = \sum_{i=1}^{M_{Rx}M_{Tx}} \lambda_i^2. \tag{6.8}$$

Inserting (6.7) and (6.8) into (6.5), we get

$$\Psi(\mathbf{R}_H) = \frac{\left(\sum_{i=1}^{M_{Rx}M_{Tx}} \lambda_i \right)^2}{\sum_{i=1}^{M_{Rx}M_{Tx}} \lambda_i^2}. \tag{6.9}$$

Higher diversity equates with a more uniform power distribution across eigenvalues. As expected, from (6.9), this translates to a higher $\Psi(\mathbf{R}_H)$.

Closely related to the diversity measure is the *correlation metric*, which quantifies the correlation between paths in a MIMO channel. Given $\Psi(\mathbf{R}_H)$, the correlation measure $\Phi(\mathbf{R}_H)$ is computed as (127)

$$\Phi(\mathbf{R}_H) = \sqrt{\frac{1 - M_{Rx}M_{Tx}/\Psi(\mathbf{R}_H)}{1 - M_{Rx}M_{Tx}}}. \tag{6.10}$$

Substituting (6.6) into (6.10), we see that $\Phi(\mathbf{R}_H)$ is bounded by $0 \leq \Phi(\mathbf{R}_H) \leq 1$.

6.1.3 The Demmel Condition Number

The Demmel condition number is a measure of the "invertability" of the channel matrix $\mathbf{H}$. It was first introduced by Demmel in Ref. 129. Kyösti et al. used it in the validation of the *wireless world initiative new radio* (WINNER) channel model (130). Czink also uses it in the validation of the RCM (80).

The Demmel condition number is the ratio of the total energy in **H** to its smallest singular value, viz.,

$$\kappa_D = \frac{\|\mathbf{H}\|_F}{\sigma_{H,\,\text{min}}}, \tag{6.11}$$

where $\sigma_{H,\text{min}}$ is the smallest singular value of **H**. It is important to note that for measured data, the smallest singular value is often determined by the noise floor of the receiver. In this instance, the Demmel condition number is not applicable. It is, however, applicable to noiseless simulated data, which, in a way, leads to practical relevance.

6.1.4 The Environmental Characterization Metric

The *environmental characterization metric* was first proposed in Refs. 80 and 131. It quantifies the *spread* of multipath components in a given channel, where *spread* refers to their average spread in the double-directional angle-delay space. The environmental characterization metric is therefore suitable for cluster-based models, where the multipath component parameters are known. One of the unique features of the environmental characterization metric is that it is computed directly from the *multipath component* parameters (i.e., not from the H-matrix) and is thus system independent.

The methodology is as follows. A super-resolution algorithm is used to obtain the multipath component parameters directly from the data. The environmental characterization metric is then computed from these parameters. This is used as the baseline. The model under test is then used to generate a synthetic data set. From this, a second environmental characterization metric is computed and compared with that first.

The environmental characterization metric $\mathbf{C}_\pi$ is computed as (131)

$$\mathbf{C}_\pi = \frac{\sum_{\ell=1}^{L} |\gamma_l|^2 (\boldsymbol{\pi}_\ell - \overline{\boldsymbol{\pi}})(\boldsymbol{\pi}_\ell - \overline{\boldsymbol{\pi}})^T}{\sum_{\ell=1}^{L} |\gamma_\ell|^2}, \tag{6.12}$$

where γ_l are the $\ell \in \{1, \ldots, L\}$ complex path gains, $\boldsymbol{\pi}_\ell$ is vector representing the normalized path parameters, and $\overline{\boldsymbol{\pi}}$ are the average normalized path parameters. For each path,

$$\boldsymbol{\pi}_\ell = \begin{bmatrix} x_{\text{Rx},\ell} & y_{\text{Rx},\ell} & z_{\text{Rx},\ell} & x_{\text{Tx},\ell} & y_{\text{Tx},\ell} & z_{\text{Tx},\ell} & \overline{\tau}_\ell \end{bmatrix}, \tag{6.13}$$

where $(x_{X,\ell}, y_{X,\ell}, z_{X,\ell})$ represents the Cartesian-coordinate transform of the angular path parameters for $X \in \{\text{Rx}, \text{Tx}\}$, and

$$\overline{\tau}_\ell = \frac{\tau_\ell}{\max_\ell(\tau_\ell)} \tag{6.14}$$

is the normalized path delay. The angular parameters are transformed to Cartesian coordinates according to

$$\begin{bmatrix} x_{X,\ell} \\ y_{X,\ell} \\ z_{X,\ell} \end{bmatrix} = \frac{1}{2} \begin{bmatrix} \sin(\varphi_{X,\ell})\sin(\theta_{X,\ell}) \\ \sin(\varphi_{X,\ell})\cos(\theta_{X,\ell}) \\ \cos(\theta_{X,\ell}) \end{bmatrix}, \tag{6.15}$$

where $\varphi_{X,\ell}$ and $\theta_{X,\ell}$ are the azimuth and elevation angles of the ℓth multipath component. The angular parameters are normalized such that the maximum Euclidean distance between any two paths is equal to or less than unity. The average parameter vector $\overline{\pi}$ is computed as

$$\overline{\pi} = \frac{\sum_{\ell=1}^{L}|\gamma_\ell|^2 \pi_l}{\sum_{\ell=1}^{L}|\gamma_\ell|^2}. \tag{6.16}$$

The environmental characterization metric gives some idea of how "spread" the multipath component parameters are. Through normalization, different scenarios can be compared fairly. The assumption is that the more "spread" a channel, the higher the diversity. The singular values of $\mathbf{C}_\pi$ are important to characterizing the spatial structure of the channel; the largest singular value indicates the largest spreading dimension. The *distance* between singular values is an indication of how spread the parameters are. Plotting the cumulative distribution function (CDF) of the singular values indicates their variability. The diagonal values of $\mathbf{C}_\pi$ indicate the spread of each of the normalized angular parameters ($x_{X,\ell}$, $y_{X,\ell}$, $z_{X,\ell}$) and the root mean-squared (RMS) delay spread. Its trace, $\mathrm{tr}(\mathbf{C}_\pi) = \sum_i \lambda_{C,i}$, is dominated by the large singular values of $\mathbf{C}_\pi$.

Example: The Random Cluster Model Here, we present several experimental results using the RCM (Chapter 4) and real-life data. The following briefly describes the experimental setup (132). The data were gathered in a non–line-of-sight (NLoS) walled office environment using the Electrobit Propsound channel sounder (124). The sounder is described in more detail in Ref. 133. The Propsound channel sounder is an 8×56 element switched array sounder. For the data used here, the transmitter used a 28-element dual polarized array, and the receiver had an 8-element monopole array. The transmitter broadcast 255 chip binary phase-shift keyed maximal-length sequences (BPSK ML sequences) at 100 Mchips/s, at a carrier frequency of 2.55 GHz. The sounder is capable of switching through all receive antennas every $t_{\mathrm{Rx,scan}} = 1542.24$ μs. Because of data transfer delays, the sounder obtained one complete MIMO response every 10.8 ms on average. The transmitter was mobile, whereas the receiver remained fixed. Once the data were recorded, the *initialization and searching improved space-alternating generalized expectation maximization* (ISIS) algorithm was used to compute the multipath component parameters (95). Details on the validation procedure can be found in Ref. 80.

Figure 6.1a plots the CDFs of the first five singular values of the environmental characterization metric. In general, the RCM does a good job of representing

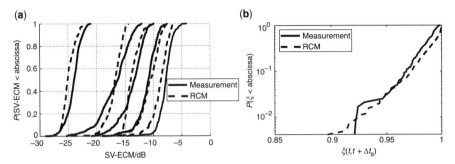

Figure 6.1 (a) The first five singular values of the environmental characterization metric (SV-ECM), and (b) the CDF of the distances between ECM singular values, computed from all snapshots, using adjacent snapshots in time.

the multipath structure of the channel, as the singular values of the environmental characterization metric predicted by the RCM closely follow those of the channel.

The environmental characterization metric can also be used to evaluate the RCM's ability to model time-variation. One method is to look at the distance between the singular values as they progress in time. Let $\xi(t, t + \Delta t_s)$ denote the distances between all singular values of the environmental characterization metric, computed between two adjacent snapshots in time, where Δt_s is the snapshot period. Figure 6.1b is a plot of the CDF of $\xi(t, t + \Delta t_s)$, where the abscissa denotes the largest to smallest changes in the distance between singular values, as we move left to right. The left side of the CDF is more jagged because there were fewer large changes in the singular value distances than there were smaller changes. For the majority of the data set, the RCM closely follows the measured data.

6.1.5 Correlation Matrix Difference Metric

As discussed in Chapter 2, one of the goals of correlative models is to approximate the full correlation $\mathbf{R}_H$ as closely as possible. In Ref. 36, Yu et al. define the model error as the difference between the correlation matrix $\mathbf{R}_H$, computed from the channel data, and the model correlation matrix $\mathbf{R}_{\text{model}}$. The error $\psi(\mathbf{R}_H, \mathbf{R}_{\text{model}})$ can be expressed as

$$\psi(\mathbf{R}_H, \mathbf{R}_{\text{model}}) = \frac{\|\mathbf{R}_H - \mathbf{R}_{\text{model}}\|_F}{\|\mathbf{R}_{\text{model}}\|_F}. \tag{6.17}$$

where $\|\cdot\|_F$ is the Frobenius norm. Whereas this metric is intuitively satisfying, in practice, it tends to overstate the model error for small differences in $\mathbf{R}_{\text{model}}$. For example, a large $\psi(\mathbf{R}_H, \mathbf{R}_{\text{model}})$ does not necessarily lead to poor performance in other metrics, such as MIMO capacity. Equation (6.17) can be extended to the wideband case by averaging the difference across all delay taps (36).

6.2 WMSDR EXPERIMENTAL SETUP

Here, we describe the WMSDR experimental setup, including the conditions under which the data were gathered.

6.2.1 Terminology

First, at the transmitter, a *frame* of data is generated. This frame is repeated continuously over the air, without any gaps between frames. Because of hardware limitations, the receiver PC does not work in real time when recording data. Rather, it gathers data for some short period of time and then goes offline to record the data to disk. This cycle is repeated for the duration of the measurement, or until the receiver disk is full. This process is illustrated in Fig. 6.2. We refer to a period where the receiver PC is online as a *snapshot*. Typically, the snapshot period is designed to be large enough to include several frames. The total measurement time is defined by the number of snapshots, plus the gaps in between.

6.2.2 Measurement Description

For the data considered in this chapter, the transmitter was configured to broadcast four independent ML sequences of 4095 chips continuously at 4 Mchips/s using a BPSK constellation. This means that the period of each frame was 1 ms. Using the MIMO matched filtering technique, one impulse estimate for each frame of data, or every 1 ms, was computed. The ML-sequence length was long enough to ensure good receive SNR (>85 dB) but was short enough so that the channel did not change appreciably over the course of a frame. For fixed measurements, the channel coherence time is typically much larger than 1 ms (for example, see Ref. 134).

At the receiver, we recorded samples at 8 MSamples/s, for approximately 8 ms continuous recording time per snapshot. This means that each snapshot contains at

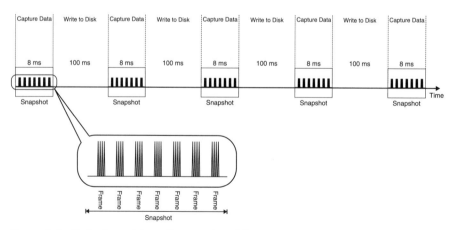

Figure 6.2 Illustration of how the receiver PC gathers data. Because of hardware limitations, the receiver PC cannot work continuously.

Figure 6.3 WMSDR outdoor measurement environment.

least seven frames. At each receive location, five snapshots were taken with about 100 ms between snapshots, when the receiver PC went offline. This means that, for each location, there were 35 H-matrices over approximately 440 ms. The distance between the transmitter and receiver ranged from approximately 10 m to 100 m. A total of 525 H-matrices were recorded over 15 locations. Both the transmitter and receiver work at a center frequency of 2135 MHz. The receiver sample rate was chosen such that the transmit signal was oversampled by two times. This reduced the noise floor and relaxed the anti-aliasing filter design requirements.

All measurement locations were in an open parking lot, with trees, cars, and buildings comprising the majority of the scatterers within 300 m. Figure 6.3 shows a picture of the parking lot and surrounding area. For all but two locations, the receiver had a direct line-of-sight to the transmitter. When the transmitter was shadowed from the receiver, the receiver was able to resolve multipath energy up to the fourth delay tap. This means that the receiver could resolve multipath energy from scatterers as far as 300 m.

The test environment and equipment setup here were chosen specifically to test one extreme of the wideband MIMO channel. The relatively short propagation distance in a sparse scattering environment (a parking lot) combined with a directive antenna array at the transmitter meant that the measured channel contained relatively few scatterers. In addition, the automatic gain control (AGC) at the receiver forced it to focus on the line-of-sight path, and the contributions of most scatterers fell below the noise floor. This, in turn, meant that the measured channel offered relatively low diversity. Using data collected from this environment tests a given model's ability to approximate the spatial structure of low-diversity channels.

6.3 BYU WIDEBAND CHANNEL SOUNDER EXPERIMENTAL SETUP

6.3.1 BYU Transmitter Set

The BYU wideband transmitter functional diagram is shown in Fig. 6.4. The main components of the transmitter consist of a vector signal generator, rubidium clock source, an electronic 1 : 8 RF switch, a custom timing and synchronization unit, and associated antennas and amplifiers.

The vector signal generator is capable of generating many different waveforms. For the purposes of sounding the wideband MIMO channel, the signal generator was programmed to generate 80 discrete pilot tones at 1 MHz spacing, for a total effective sounding bandwidth of 80 MHz. The output of the signal generator is connected to an electronic 1 : 8 RF switch that distributes the RF signal to one of eight antennas.

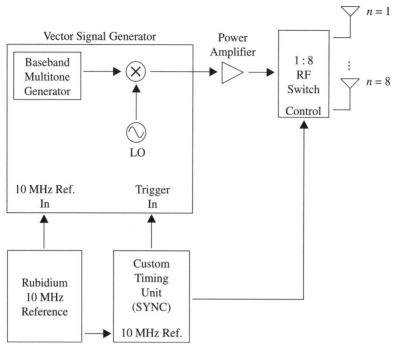

Figure 6.4 BYU channel sounder transmitter block diagram.

The transmitter is capable of generating a maximum of 27 dBm per antenna. For the measurements considered here, it was set to 23 dBm/antenna.

The RF switch position is determined by the synchronization unit. There is an identical synchronization unit at both transmitter and receiver. Both synchronization units are synchronized to a rubidium timing reference. Before each data run, both synchronization units are synchronized to each other to minimize timing drift over the course of the measurement. It is assumed that the rubidium sources are precise enough to keep the synchronization units synchronized for the duration of the measurement. All RF sources are locked to the rubidium source, which minimizes relative frequency drift. For the measurements below, the synchronization unit at the transmitter was programmed to excite each antenna for $t_{\text{Tx,dwell}} = 50\ \mu\text{s}$. This means that at least $t_{\text{Tx,scan}} = 50 \times 8 = 400\ \mu\text{s}$ was needed to scan the entire array. The synchronization unit at the transmitter also ensures that the signal source sequence starts at the proper time instant for each antenna.

6.3.2 BYU Receiver Set

The receiver functional diagram is shown in Fig. 6.5. It consists of an electronic $8:1$ RF switch, a PC equipped with a data acquisition card, synchronization unit, a rubidium clock source, and all associated RF hardware including amplifiers, antennas,

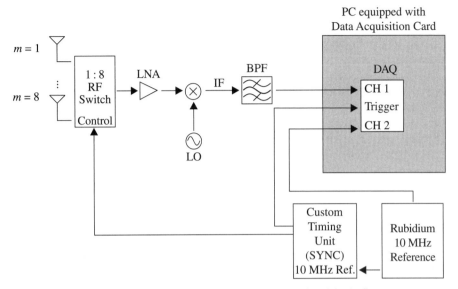

Figure 6.5 BYU wideband channel sounder receiver block diagram.

LO, RF mixer, and bandpass filter. The output of the RF switch is amplified by 40 dB, down-converted to a 50 MHz *intermediate frequency* (IF), and fed to one channel of the data acquisition card. The down-conversion shifts the channel information so that it occupies 10–90 MHz at IF and brings it within the data acquisition card sample range. The data acquisition card samples the signal at 500 MSamples/s, which means that it captures 0–250 MHz. This includes the entire 80 MHz channel.

The 10 MHz signal from the rubidium source is also sampled to ensure phase coherence with the transmitter in postprocessing. Each receive-antenna is sampled for $t_{\text{Rx,dwell}} = t_{\text{Tx,scan}} = 400\ \mu\text{s}$, the full cycle time of the transmitter. This means that it takes at least $t_{\text{Rx,scan}} = 3.2$ ms to measure a complete MIMO channel response. Because of hardware limitations, there is a gap between each complete channel response measurement. For the data below, the receiver was programmed to measure a complete response every 25.6 ms and recorded data for two different durations, depending on the location. For locations 1–3 and 7–8, approximately 15 s of data were recorded. For locations 4–6, approximately 10 s of data were recorded.

6.3.3 Measurement Description

The BYU data at each location were recorded using two different carrier frequencies, 2.55 GHz and 5.2 GHz. Both transmitter and receiver employed a circular array of vertically polarized monopoles set to approximately $\lambda/2$ spacing for each center frequency. Both the transmitter and receiver were located indoors, most often in separate rooms. The transmitter was fixed to one of two positions, and the receiver was moved to one of eight different locations. All measurements were done in a fixed point-to-point environment with no line-of-sight.

Compared with the WMSDR setup described above, the BYU data were collected in a much more scatter-rich environment. Omnidirectional antennas at both link ends ensure scatterers in the entire azimuth plane are well illuminated. The indoor environment provides numerous scatterers in the form of walls, floors, furniture, and so forth. In addition, the NLoS environment ensures that the receiver focuses on weaker reflected paths, which, in turn, increases diversity.

The increased effective sounding bandwidth versus the WMSDR also means that there is better scatterer resolution with respect to propagation distance. For example, for the WMSDR, the 2 MHz effective bandwidth means that all energy within 500 ns is grouped into one delay bin. Referring to the oval of scatterers model shown in Chapter 2, this means that each annulus is approximately 150 m bigger than the last. On the other hand, the BYU sounder's 80 MHz bandwidth translates to an approximate scatter resolution of 3.75 m. The increased resolution translates to a larger number of resolved scatterers and thus a much more scatter-rich channel. The BYU data set was chosen specifically to test the structured model's ability to model scatter-rich environments.

6.4 EXPERIMENTAL RESULTS

In this section, we are interested in verifying the performance of the structured model using real-life data. Specifically, we are interested in the structured model's ability to predict the true *capacity* of the channel and to model its spatial structure. The first criterion leads to a quantitative metric; the difference between the true channel capacity and that predicted by the structured model can be computed. These results can be compared with those obtained using the wideband Kronecker model.

Both the structured and Kronecker models are correlative channel models. As such, it is important to somehow test their ability to model the spatial structure of real-life channels. To this end, the MIMO APS of the channel is compared with that computed using the structured and wideband Kronecker models. This leads to a qualitative measure; the location and relative strength of peaks in the MIMO APS can be compared to determine how well each model fairs in approximating the spatial structure of the channel.

One of the unique features of the structured model is that it models the correlation between delays, as well as across transmitters and receivers. The last section of the chapter attempts to answer the question, "How important is it to model the correlation between channels at different delays?" The wideband correlation matrix $\mathbf{R}_{\mathrm{WB},H}$ is structured such that elements that describe the correlation between channels at different delays are contained in the off-diagonal blocks of the matrix. The "energy" contained in these elements can be quantified and compared between measurements. The assertion is that the more energy contained in these off-diagonal blocks, the more important it is to model these correlations to better approximate the spatial structure of the channel.

6.4.1 Capacity Measure: Methodology

The MIMO literature on channel modeling often deals with the average or *ergodic* capacity and its variability. The variability is measured using the *capacity CDF*. Here, we are primarily interested in comparing the structured model's ability to predict the capacity of measured channels with that of the wideband Kronecker model. We also compare both models to the true capacity of the channel, which gives us an absolute reference.

First, the measured data are used to compute the parameters of each model. Using these parameters, an ensemble of H-tensors is generated using both models. From this, the ergodic capacity is computed. We compare this with the true ergodic capacity of the channel. Given the ergodic capacity at all locations, we can plot a capacity CDF for each model and compare them with the true capacity of the channel. The following describes this process in more detail.

6.4.1.1 *Generating H-tensors from Channel Models*   The H-tensors for each location were computed using the WMSDR and BYU data. For the WMSDR data, all 15 locations for all times were considered. The data were divided by location. For the BYU data, the first 1.28 s of each data set for each location was considered. For each location, data were recorded at two different times and at two different frequencies. Thus, the data for each location were divided into four subsets; one collected at 2.55 GHz, one collected at 5.2 GHz, and at two different times per center frequency.

For convenience, the following repeats key equations from Chapter 3. After computing $\mathcal{H}$ from the recorded data, we computed $\mathbf{R}_{\mathrm{WB},H}$ using

$$\mathbf{R}_{\mathrm{WB},H} = \mathrm{E}\{\mathrm{vec}(\mathcal{H})\mathrm{vec}^H(\mathcal{H})\}. \tag{6.18}$$

An ensemble of synthetic H-tensors $\mathcal{H}_{\mathrm{synth}}$ is then computed using

$$\mathcal{H}_{\mathrm{synth}} = \mathrm{unvec}(\mathbf{R}_{\mathrm{WB},H}^{1/2}\mathbf{g}_{\mathcal{H}}), \tag{6.19}$$

where $\mathbf{g}_H \in \mathbb{C}^{M_{\mathrm{Rx}}M_{\mathrm{Tx}}D \times 1}$ is a complex Gaussian vector. $\mathcal{H}_{\mathrm{synth}}$ is considered to be the basis for comparison. Computing $\mathcal{H}_{\mathrm{synth}}$ in this way removes any non-Rayleigh statistics from the data while maintaining the spatial characteristics of the measured channel. By nature of the Rayleigh assumption implicit in our model, the goal is to reproduce the spatial characteristics of the measured channel, not necessarily its statistics. This method was used, for example, by Weichselberger in Ref. 51.

In the case of the structured model, $\mathcal{H}$ is used to compute the one-sided correlation matrices $\mathbf{R}_{\mathrm{Rx}}$, $\mathbf{R}_{\mathrm{Tx}}$, and $\mathbf{R}_{\mathrm{Del}}$, viz.,

$$\mathbf{R}_{\mathrm{Rx}} = \mathrm{E}\{\mathbf{H}_{(1)}\mathbf{H}_{(1)}^H\}, \tag{6.20}$$

where $\mathbf{H}_{(n)}$ is the nth matrix unfolding of $\mathcal{H}$. The eigenvalue decomposition (EVD) is then used to compute the one-sided eigenbases $\mathbf{U}_{\text{Rx}}$, $\mathbf{U}_{\text{Tx}}$, and $\mathbf{U}_{\text{Del}}$, viz.,

$$\mathbf{R}_{\text{Rx}} = \sum_{i=1}^{M_{\text{Rx}}} \lambda_{\text{Rx},i} \mathbf{u}_{\text{Rx},i} \mathbf{u}_{\text{Rx},i}^H = \mathbf{U}_{\text{Rx}} \mathbf{\Lambda}_{\text{Rx}} \mathbf{U}_{\text{Rx}}^T$$

$$\mathbf{R}_{\text{Tx}} = \sum_{j=1}^{M_{\text{Tx}}} \lambda_{\text{Tx},j} \mathbf{u}_{\text{Tx},j} \mathbf{u}_{\text{Tx},j}^H = \mathbf{U}_{\text{Tx}} \mathbf{\Lambda}_{\text{Rx}} \mathbf{U}_{\text{Tx}}^H \qquad (6.21)$$

$$\mathbf{R}_{\text{Del}} = \sum_{k=1}^{D} \lambda_{\text{Del},k} \mathbf{u}_{\text{Del},k} \mathbf{u}_{\text{Del},k}^H = \mathbf{U}_{\text{Del}} \mathbf{\Lambda}_{\text{Del}} \mathbf{U}_{\text{Del}}^H$$

Next, the wideband coupling coefficients ω_{ijk} for $i = \{1, \ldots, M_{\text{Rx}}\}, j = \{1, \ldots, M_{\text{Tx}}\}$, and $k = \{1, \ldots, D\}$ are computed using

$$\omega_{ijk} = (\mathbf{u}_{\text{Del},k} \otimes \mathbf{u}_{\text{Tx},j} \otimes \mathbf{u}_{\text{Rx},i})^H \mathbf{R}_{\text{WB},H} (\mathbf{u}_{\text{Del},k} \otimes \mathbf{u}_{\text{Tx},j} \otimes \mathbf{u}_{\text{Rx},i}). \qquad (6.22)$$

Finally, the structured model synthesis equation is used to generate an ensemble of $\mathcal{H}_{\text{struct}}$,

$$\mathcal{H}_{\text{struct}} = \mathcal{G} \times_1 \mathbf{U}_{\text{Rx}} \times_2 \mathbf{U}_{\text{Tx}} \times_3 \mathbf{U}_{\text{Del}}, \qquad (6.23)$$

where $\mathcal{G} \in \mathbb{C}^{M_{\text{Rx}} \times M_{\text{Tx}} \times D}$ is the tensor whose elements $w_{mnd} = g_{mnd} \sqrt{\omega_{mnd}}$, and g_{mnd} is a complex-Gaussian random variable.

For comparison, an ensemble of H-tensors is computed using the wideband Kronecker model. First, the one-sided correlation matrices $\mathbf{R}_{\text{Rx}}[d]$ and $\mathbf{R}_{\text{Tx}}[d]$ at each delay tap are computed, where

$$\mathbf{R}_{\text{Rx}}[d] = E\{\mathbf{H}[d]\mathbf{H}^H[d]\}$$
$$\mathbf{R}_{\text{Tx}}[d] = E\{\mathbf{H}^T[d]\mathbf{H}^*[d]\}, \qquad (6.24)$$

and $\mathbf{H}[d]$ is the $M_{\text{Rx}} \times M_{\text{Tx}}$ formed by taking all h_{mnd} for a fixed d. The wideband Kronecker model synthesis equation is used to compute an ensemble of $\mathbf{H}_{\text{Kron}}[d]$ at each delay, viz.,

$$\mathbf{H}_{\text{Kron}}[d] = \mathbf{R}_{\text{Rx}}^{1/2}[d] \mathbf{G} (\mathbf{R}_{\text{Tx}}^{1/2}[d])^T, \qquad (6.25)$$

where $\mathbf{G} \in \mathbb{C}^{M_{\text{Rx}} \times M_{\text{Tx}}}$ is a spatially white matrix with complex-Gaussian entries. The elements from $\mathbf{H}_{\text{Kron}}[d]$ for $d = \{1, \ldots, D\}$ can be directly mapped to the H-tensor $\mathcal{H}_{\text{kron}}$.

6.4.1.2 A Note on the Selection of Maximum Delay

For practical purposes, D must be fixed for any given measurement. The *power delay profiles* (PDPs) computed from the WMSDR data were fairly uniform in terms of the excess delay. The average PDP never displayed energy beyond the fourth delay bin. Therefore, in the following analysis, the WMSDR data set is restricted to the

$(M_{Rx}, M_{Tx}, D) = (4, 4, 4)$ case, over all 15 locations. Selecting $D = 4$ ensures that all PDPs represent the energy of the channel, down to the noise floor of the receiver.

On the other hand, the PDPs computed from the BYU data were much more dynamic. The best way to quantify the average PDP is via the average RMS delay spread. The RMS delay spread [computed using (12) and (13) from Ref. 135], averaged over all locations and all times was computed to be approximately 56 ns at -120 dB threshold. The BYU data were divided into different sets by considering subsets of the data for which $D \in \{4, \ldots, 10\}$. This corresponds with delays of 50–125 ns. At $D = 4$, most of the channel energy is captured in the PDP. At $D = 10$, energy well past the RMS delay is captured in the PDP. In any case, the truncated PDPs can be viewed as approximations of the real-life channel, with the approximation getting better as D gets larger.

6.4.1.3 *Capacity Results: WMSDR Data Set*   Table 6.1 summarizes the capacity results for the case when $(M_{Rx}, M_{Tx}, D) = (4, 4, 4)$ averaging over an ensemble of 10,000 iterations. The number of parameters needed to compute $\mathcal{H}_{synth}$, $\mathcal{H}_{Kron}$, and $\mathcal{H}_{struct}$ are listed in Table 6.2. Figure 6.6 shows the modeled versus measured capacity for the structured model and the wideband Kronecker model. The diagonal represents the case of no model error. Model error is defined as

$$\% \text{ Error} = \frac{C_X - C_{synth}}{C_{synth}} \times 100, \tag{6.26}$$

where $X \in \{\text{Kron, struct}\}$.

Table 6.1 Kronecker versus structured model error

Location	C_{synth} (bps)	C_{Kron} (bps)	C_{struct} (bps)	% Error Kronecker	% Error Structured
0901	16.7	9.6	16.3	42.5	2.4
1100	15.0	10.4	15.3	30.7	2.0
1215	17.7	11.3	17.4	36.2	1.7
1354	17.1	8.7	16.9	49.1	1.2
1750	17.1	12.2	16.5	28.7	3.5
1844	19.7	11.7	18.7	40.6	5.1
1931	17.0	10.9	16.0	35.9	5.9
2021	16.6	10.7	15.9	35.5	4.2
2116	16.6	11.1	16.6	33.1	0
2537	14.9	10.8	15.3	27.5	2.7
2652	11.2	8.8	12.4	21.4	10.7
2744	12.6	9.8	14.0	22.2	11.1
2841	15.8	11.6	15.0	26.6	5.1
3108	13.7	8.7	14.9	36.5	8.8
3223	15.7	9.0	15.9	42.7	1.3
				Average	
				33.9%	**4.4%**

Table 6.2 Number of parameters required to compute an exemplar H-tensor for the three models considered above

Model	Number of Required Parameters (M_{Rx}, M_{Tx}, D)	Number of Required Parameters $(4, 4, 4)$
$\mathcal{H}_{synth}$	$(M_{Rx}M_{Tx}D)^2$	4096
$\mathcal{H}_{Kron}$	$D(M_{Rx}^2 + M_{Tx}^2)$	128
$\mathcal{H}_{struct}$	$(M_{Rx}M_{Tx}D) + (M_{Rx}^2 + M_{Tx}^2 + D^2)$	112

Discussion   In general, the structured model performed very well, especially when compared with the wideband Kronecker model. For all locations, the structured model error is 4.4%, whereas that of the wideband Kronecker model is 33.9%. In Fig. 6.6, the Kronecker model consistently underestimates the capacity of the channel. These results are in line with those given by Yu in Ref. 73.

6.4.1.4 Capacity Results: BYU Data Set

The BYU data are divided into two subsets, one recorded at 2.55 GHz (Data 255) and the second at 5.2 GHz (Data 52). Each subset contains data gathered at eight different locations at two different times per location. Table 6.3 and Table 6.4 list the results for the average capacity error over all locations for Data 255 and Data 52, respectively. For each data set, subsets of the data were considered, when $M_{Rx} = M_{Tx} = 4, 6$, and 8, and $D = \{4, \ldots, 10\}$. Table 6.5 lists the number of parameters used to compute $\mathcal{H}_{synth}$, $\mathcal{H}_{kron}$, and $\mathcal{H}_{struct}$. This shows that, for all cases considered here, *fewer* parameters were required to generate an exemplar H-tensor using the structured model versus the wideband Kronecker model. Figure 6.7a shows a modeled versus measured capacity graph for

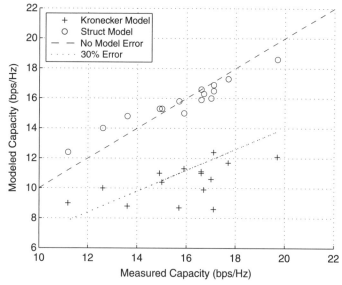

Figure 6.6 Modeled versus measured capacity for WMSDR data, all locations.

Table 6.3 Model error for Data 255

M_{Rx}, M_{Tx}, D	% Error Kronecker	% Error Structured
4, 4, 4	42.6	4.8
4, 4, 5	10.9	4.8
4, 4, 6	38.8	5.2
4, 4, 7	37.1	4.5
4, 4, 8	36.5	4.3
4, 4, 9	35.6	5.1
4, 4, 10	35.5	4.6
6, 6, 4	52.3	4.6
6, 6, 5	51.4	3.4
6, 6, 6	49.4	3.4
6, 6, 7	48.6	3.4
6, 6, 8	47.8	3.6
6, 6, 9	47.2	3.4
6, 6, 10	46.9	3.2
8, 8, 4	60.1	6.1
8, 8, 5	58.1	5.3
8, 8, 6	55.9	5.2
8, 8, 7	55.1	4.3
8, 8, 8	54.6	4.3
8, 8, 9	54.3	4.1
8, 8, 10	53.8	4.4
Average		
	47.7	**4.4**

all locations, all data sets, and all cases listed above. Figure 6.7b shows the capacity CDF for all locations, all data sets, and all cases. The results are discussed in the following section.

Discussion   In light of the theory presented in Chapter 3, one would expect the Kronecker model to exhibit increased modeling error as the number of antennas increases. From Refs. 51 and 73, the number of artifact paths resulting from the Kronecker assumption increases as the spatial resolution increases. Tables 6.3 and 6.4 show that this is indeed the case. On the other hand, the structured model error remains relatively constant regardless of array size. Table 6.6 shows the average model error, broken down for the cases when $M_{Rx} = M_{Tx} = 4$, 6, and 8.

As with the WMSDR data, the structured model performed quite well overall. The total error, averaging over all data sets and all cases mentioned above, was 46.1% for the wideband Kronecker model and 4.1% for the structured model. This is close to the error reported in the WMSDR case, which was 4.4%.

As illustrated in Fig. 6.7b, the capacity CDF from the structured model followed the true CDF much more closely than did the Kronecker model. In this plot, the Kronecker model consistently underestimates the true capacity.

Table 6.4 Model error for Data 52

M_{Rx}, M_{Tx}, D	% Error Kronecker	% Error Structured
4, 4, 4	39.8	3.8
4, 4, 5	38.0	2.9
4, 4, 6	35.3	4.4
4, 4, 7	35.5	4.5
4, 4, 8	35.1	3.4
4, 4, 9	34.7	4.1
4, 4, 10	34.7	3.9
6, 6, 4	51.1	4.7
6, 6, 5	49.7	4.0
6, 6, 6	47.4	4.0
6, 6, 7	47.1	3.5
6, 6, 8	46.1	3.1
6, 6, 9	45.0	3.3
6, 6, 10	45.1	2.9
8, 8, 4	58.2	5.2
8, 8, 5	56.7	4.0
8, 8, 6	54.6	3.8
8, 8, 7	54.2	3.3
8, 8, 8	53.7	3.3
8, 8, 9	53.0	3.8
8, 8, 10	53.0	3.2
Average		
	44.4	**3.7**

6.4.2 Results: MIMO APS and Spatial Structure

Using $\mathcal{H}$, $\mathcal{H}_{\mathrm{Kron}}$, and $\mathcal{H}_{\mathrm{struct}}$ computed from the BYU data, the correlation matrices $\mathbf{R}_{\mathrm{WB},H[d]H[d]}$, $\mathbf{R}_{\mathrm{WB},H[d]H[d],\mathrm{Kron}}$, and $\mathbf{R}_{\mathrm{WB},H[d]H[d],\mathrm{struct}}$ were computed for each delay d. Recall that $\mathbf{R}_{\mathrm{WB},H[d]H[d]}$ is formed from the elements of $\mathbf{R}_{\mathrm{WB},H[d]H[d]}$ such that

$$
\mathbf{R}_{\mathrm{WB},H} =
\begin{pmatrix}
E\{\mathrm{vec}(\mathbf{H}[1])\mathrm{vec}^H(\mathbf{H}[1])\} & \cdots & E\{\mathrm{vec}(\mathbf{H}[1])\mathrm{vec}^H(\mathbf{H}[D])\} \\
\vdots & \ddots & \vdots \\
E\{\mathrm{vec}(\mathbf{H}[D])\mathrm{vec}^H(\mathbf{H}[1])\} & \cdots & E\{\mathrm{vec}(\mathbf{H}[D])\mathrm{vec}^H(\mathbf{H}[D])\}
\end{pmatrix}
$$

$$
=
\begin{pmatrix}
\mathbf{R}_{\mathrm{WB},H[1]H[1]} & \cdots & \mathbf{R}_{\mathrm{WB},H[1]H[D]} \\
\vdots & \ddots & \vdots \\
\mathbf{R}_{\mathrm{WB},H[D]H[1]} & \cdots & \mathbf{R}_{\mathrm{WB},H[D]H[D]}
\end{pmatrix}. \tag{6.27}
$$

Table 6.5 Number of parameters for each BYU subset

M_{Rx}, M_{Tx}, D	Number of Parameters Synthetic	Number of Parameters Kronecker	Number of Parameters Structured
4, 4, 4	4096	128	112
4, 4, 5	6400	160	137
4, 4, 6	9216	192	164
4, 4, 7	12544	224	193
4, 4, 8	16384	256	224
4, 4, 9	20736	288	257
4, 4, 10	25600	320	290
6, 6, 4	20736	288	232
6, 6, 5	32400	360	277
6, 6, 6	46656	432	324
6, 6, 7	63504	504	373
6, 6, 8	82944	576	424
6, 6, 9	104976	648	477
6, 6, 10	129600	720	532
8, 8, 4	65536	512	400
8, 8, 5	102400	640	437
8, 8, 6	147456	768	548
8, 8, 7	200704	896	625
8, 8, 8	262144	1024	704
8, 8, 9	331776	1152	785
8, 8, 10	409600	1280	868

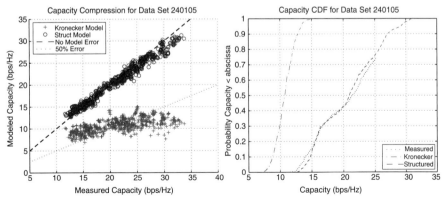

Figure 6.7 (a) Modeled versus measured capacity for BYU data, all locations, all cases. (b) Capacity CDF for BYU data, all locations, all cases.

Table 6.6 Average model error broken down by array size

M_{Rx}, M_{Tx}	% Error Kronecker	% Error Structured
4, 4	34.6	4.4
6, 6	48.3	3.6
8, 8	55.4	4.3
Total average		
	46.1	**4.1**

In a similar manner, $\mathbf{R}_{WB,H[d]H[d],Kron}$ and $\mathbf{R}_{WB,H[d]H[d],struct}$ were computed using

$$
\mathbf{R}_{WB,H[d]H[d],Kron} = E\{vec(H_{Kron}[d])vec^H(H_{Kron}[d])\}
$$
$$
\mathbf{R}_{WB,H[d]H[d],struct} = E\{vec(H_{struct}[d])vec^H(H_{struct}[d])\},
$$

(6.28)

where $H_{Kron}[d]$ and $H_{struct}[d]$ are formed using the elements of $\mathcal{H}_{Kron}$ and $\mathcal{H}_{struct}$ such that

$$
\mathbf{H}_{Kron}[d] = \begin{pmatrix} h_{Kron,11d} & \cdots & h_{Kron,1M_{Tx}d} \\ \vdots & \ddots & \vdots \\ h_{Kron,M_{Rx}1d} & \cdots & h_{Kron,M_{Rx}M_{Tx}d} \end{pmatrix}
$$
$$
\mathbf{H}_{struct}[d] = \begin{pmatrix} h_{struct,11d} & \cdots & h_{struct,1M_{Tx}d} \\ \vdots & \ddots & \vdots \\ h_{struct,M_{Rx}1d} & \cdots & h_{struct,M_{Rx}M_{Tx}d} \end{pmatrix}.
$$

(6.29)

Using the above equations, the MIMO APS can be computed for each correlation matrix at each delay. This provides a qualitative indication of how well the Kronecker and structured models represent the true spatial structure of the channel.

Figure 6.8 illustrates the MIMO APS for $\mathbf{R}_{WB,H[d]H[d]}$, $\mathbf{R}_{WB,H[d]H[d],Kron}$, and $\mathbf{R}_{WB,H[d]H[d],struct}$, computed from data at location 1, set 1b, for the first delay. The true MIMO APS, shown in Fig. 6.8a, contains many distinct resolvable paths, indicating a scatter-rich environment. Following the discussion in Chapter 3, the Kronecker APS, shown in Fig. 6.8b, emphasizes the paths closest to the AoA = AoD = 0° axes while de-emphasizing all other paths. This creates a noticeable cross-pattern. The structured model, used to produce the APS shown in Fig. 6.8c, does a much better job of representing the center of most major paths, regardless of their azimuth location.

Figure 6.9 shows the MIMO APSs for location 2, set 1b, delay bin 3. The true APS contains most of the scattered energy along the AoA = 60° and AoD = 50° axes. The Kronecker model APS contains most of its energy near the AoD = 5° axis. Again, the Kronecker APS contains a noticeable cross-shape near the central axis.

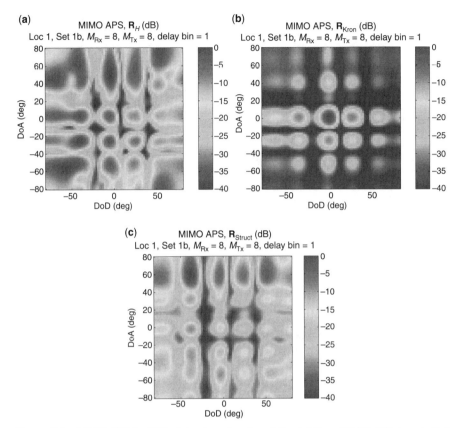

Figure 6.8 MIMO APS for BYU data, location 1, set 1b. (a) True MIMO APS computed from $\mathbf{R}_{WB,H(d)H(d)}$. (b) MIMO APS computed from $\mathbf{R}_{WB,H(d)H(d),Kron}$. (c) MIMO APS computed from $\mathbf{R}_{WB,H(d)H(d),struct}$.

The structured model APS follows the true APS more closely. It concentrates its energy in scatterers along the same AoA $= 60°$ and AoD $= 50°$ axes and approximates the centers of most scatterers.

Figure 6.10 shows the MIMO APSs for location 2, set 2b, delay bin 3. The true APS contains many small scatterers, located throughout the AoA–AoD plane, none of which occur near the center. Overall, the channel has rich scattering. The Kronecker model APS, on the other hand, focuses on a single scatterer near the center of the plane. The Kronecker model does a fairly poor job of representing the centers of any of the scatterers in the spectrum. The structured model APS, on the other hand, does a better job of representing the diversity of the channel, although some of the scatterers near the center of the spectrum are not as well represented.

The structured model did not always outperform the Kronecker model with respect to approximating the spatial structure of the channel. There were at least a few cases where the Kronecker model did a better job of modeling the spatial structure of the

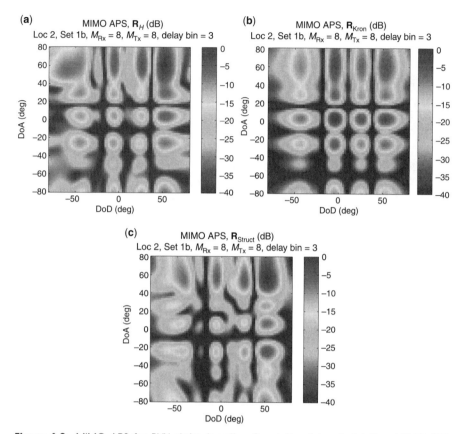

Figure 6.9 MIMO APS for BYU data, location 2, set 1b, delay 3. (a) True MIMO APS computed from $\mathbf{R}_{WB,H(d)H(d)}$. (b) MIMO APS computed from $\mathbf{R}_{WB,H(d)H(d),Kron}$. (c) MIMO APS computed from $\mathbf{R}_{WB,H(d)H(d),struct}$.

channel. This occurred when the true scatterer clusters did in fact occur at the intersection points of other scatterers, indicating that the true spectrum was separable. For example, Fig. 6.11 shows the MIMO APSs for location 3, set 1b, delay bin 2. The scatterers in the true APS follow a grid-like pattern. Also, the majority of the peaks fall at the intersection points of other major peaks. Under these circumstances, the Kronecker model does a fairly good job of representing the centers and magnitudes of most peaks. The structured model, on the other hand, seems to spread the energy throughout the spectrum. This is possibly due to the model assumption that there must be at least some coupling between scatterers.

In general, however, for most locations, the structured model followed the true MIMO APS closer than did the Kronecker model. The Kronecker model faired worse in channels where scatterers were located away from the AoA–AoD axes or had many diagonally aligned peaks. This indicated that the AoA–AoD spectra were in fact not separable the majority of the time. Thus, for the channels represented by

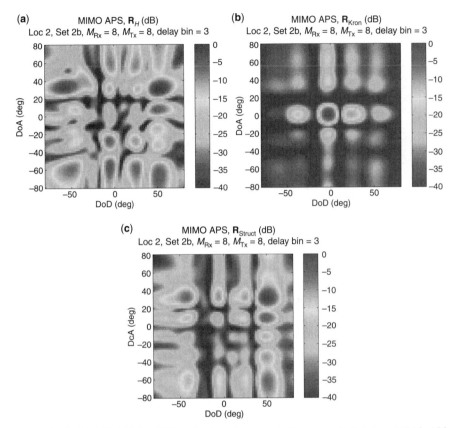

Figure 6.10 MIMO APS for BYU data, location 2, set 2b, delay 3. (a) True MIMO APS computed from $\mathbf{R}_{WB,H(d)H(d)}$. (b) MIMO APS computed from $\mathbf{R}_{WB,H(d)H(d),Kron}$. (c) MIMO APS computed from $\mathbf{R}_{WB,H(d)H(d),struct}$.

the BYU data set, the Kronecker model does not represent the spatial structure as well as the structured model.

6.4.3 Results: Wideband Correlation Matrices

Given the proposed formulation for $\mathbf{R}_{WB,H}$, we wish to quantify and visualize the correlation between scatterers located at different delay taps and compare this with the correlation between scatterers at the same delay tap. This would also qualitatively emphasize the importance of the structured model's ability to model these correlations. In the literature to date, the correlation between delays, and the structure that this would imply, has largely been ignored.

To be clear, there are *many* factors that contribute to higher correlation between channels at different delays. As we mentioned above, the WMSDR specifications are very different from those of the BYU sounder. Some factors that affect

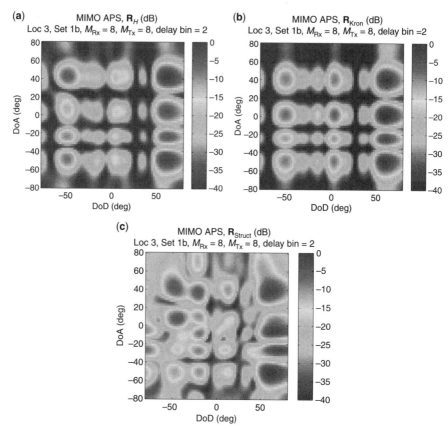

Figure 6.11 MIMO APS for BYU data, location 3, set 1b, delay 2. (a) True MIMO APS computed from $\mathbf{R}_{WB,H(d)H(d)}$. (b) MIMO APS computed from $\mathbf{R}_{WB,H(d)H(d),Kron}$. (c) MIMO APS computed from $\mathbf{R}_{WB,H(d)H(d),struct}$.

the correlation are the sounding bandwidth, which determines spatial resolution, antenna configuration, and the orientation of scatterers in the physical channel. Although it is nearly impossible to determine the contribution of each factor to the overall correlation, we hypothesize here that the increase is mainly due to two factors:

1. The increased effective sounding bandwidth (which increases spatial resolution); and

2. The increased scatter diversity (outdoor line-of-sight for the WMSDR data versus indoor non–line-of-sight for the BYU data).

The differences between the WMSDR and BYU test setup result in different correlations across delay space. Figure 6.12 shows $\mathbf{R}_{WB,H}$, computed using data from the WMSDR recorded at one location. Figure 6.13 shows $\mathbf{R}_{WB,H}$ for BYU data

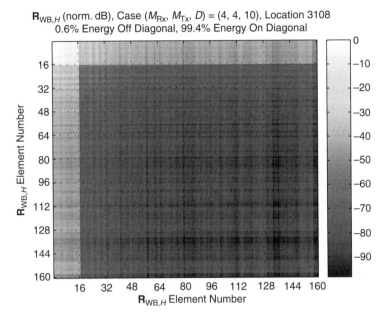

Figure 6.12 The wideband correlation matrix $\mathbf{R}_{WB,H}$ plotted for the WMSDR data for location 3108.

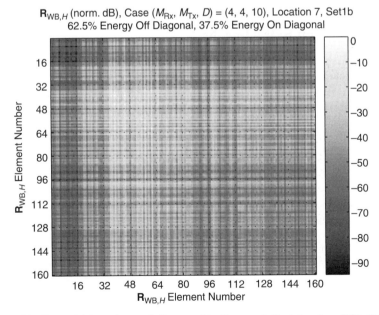

Figure 6.13 The wideband correlation matrix $\mathbf{R}_{WB,H}$ plotted for the BYU data for location 7, set 1b.

collected at a single location. From these plots, it is obvious that significantly more energy is contained in the off-diagonal blocks for the BYU data than for the WMSDR data.

The following describes a metric to quantify the increased correlation between channels at different delays. Quantitatively, we define the *weight* of each block $\mathbf{R}_{\text{WB},H[d]H[r]}$ to be

$$e_{dr} = \|\mathbf{R}_{\text{WB},H[d]H[r]}\|_F^2, \tag{6.30}$$

where the diagonal blocks are defined as those for which $r = d$, and the off-diagonal blocks are when $r \neq d$. The *total weight* is defined as

$$\varepsilon_{\text{w}} = \|\mathbf{R}_{\text{WB},H}\|_F^2. \tag{6.31}$$

Using these metrics, the total weight of all diagonal blocks can be compared with the total weight off the diagonal. For the WMSDR data in Fig. 6.12, 0.6% of the weight is contained in off-diagonal blocks, and 99.4% of it is contained on the diagonal. By contrast, for $\mathbf{R}_{\text{WB},H}$ in Fig. 6.13, 62.5% of the weight is in the off-diagonal blocks, and 37.5% is in the diagonal blocks. In general, for all locations, the BYU data contained far more weight in the off-diagonal blocks than did the WMSDR data.

6.5 DISCUSSION

6.5.1 Accuracy of the Results

At this point, we wish to emphasize the differences between the WMSDR data and the BYU data. The WMSDR is a true 4×4 wideband sounder based on a calibrated PN-sequence estimation of the channel. The BYU sounder uses the technique of spectrum sampling and switches through the entire 8×8 array to capture the MIMO response. The WMSDR has much narrower bandwidth (2 MHz effective) versus the BYU sounder (80 MHz effective). Most importantly, the data were gathered in completely different environments. The WMSDR data were collected outdoors, in a line-of-sight environment with relatively sparse scattering. The BYU data were collected indoors where no line-of-sight existed between the transmitter and receiver. Despite these differences, the structured model performs equally well with both data sets, proving the robustness of the model.

6.5.2 Sources of Error

Although the structured model performed very well overall, there was still an average error of approximately 4.5% across all data sets. This discrepancy is due to a number of factors. The following is a qualitative discussion of factors that contribute to the modeling error.

First, perhaps the largest source of error is the model assumption that the wideband correlation matrix eigenvectors $\mathbf{u}_{\mathrm{WB},k}$ can be approximated as the Kronecker product of the one-sided eigenvectors $\mathbf{u}_{\mathrm{Rx},i}$, $\mathbf{u}_{\mathrm{Tx},j}$, and $\mathbf{u}_{\mathrm{Del},k}$. This assumption reduces the number of parameters needed to estimate $\mathbf{R}_{\mathrm{WB},H}$, but obviously introduces error. From Section 6.4.2, this assumption tends to yield a "smeared" version of the original APS, as well as errors in the estimated capacity.

Another source of error is the fundamental assumption that the channel is linear. Although the wireless channel itself is governed by Maxwell's equations and is widely considered to be linear, the recorded data also includes the nonlinear effects of electronics at the transmitter output and receiver input. Devices such as DACs, ADCs, mixers, and amplifiers, all exhibit nonlinearities. Each of these devices has limits, within which their behavior is almost linear, or *quasi*-linear. Both the WMSDR and BYU channel sounder are carefully designed to work within these limits, such that the overall effect of device nonlinearities is small if not negligible. Nevertheless, these nonlinearities could affect the performance of any linear models applied to the data.

6.6 SUMMARY AND DISCUSSION

The chapter began with a summary of a few commonly used validation metrics. The ergodic capacity is the most popular metric used to validate MIMO channel models. The promise of higher bandwidth efficiencies means that it is important for a model to predict accurately the capacity of a MIMO channel. The diversity and correlation metrics are measures of how correlated the paths are in a MIMO channel. This is closely related to the distribution of the channel eigenvalues. The environmental characterization metric is useful for cluster models, in which the statistical information of all multipath components is known. Finally, the correlation matrix difference metric quantified the difference between $\mathbf{R}_H[d]$ and $\mathbf{R}_{\mathrm{model}}[d]$ directly. This is useful in evaluating the ability of correlative models to approximate the full correlation.

The latter half of the chapter dealt with comparative evaluations of the structured and Kronecker models using data from two widely different apparatuses, namely the WMSDR and BYU sounder. The WMSDR and BYU data were collected under very different conditions. Moreover, the WMSDR data were collected in an outdoor line-of-sight environment. This, in turn, meant that the channel had relatively few scatterers, and thus poor diversity. In contrast, the BYU data were gathered in a scatter-rich indoor environment with no line-of-sight between transmitter and receiver.

We then evaluated the structured model's performance using a few different metrics. The first was channel capacity. We showed the structured model could be used to accurately predict the capacity of real-life channel. The structured model performed consistently well for both the WMSDR and BYU data, over all locations, and for all cases. The structured model error was approximately 4% for both sets. From the WMSDR to the BYU data, the Kronecker model error increased from 34% to 46%. For the BYU data, we considered the cases for 4×4, 6×6, and 8×8 MIMO channels. From this analysis, we saw that the structured model error remained essentially constant with increasing array size. In contrast, the Kronecker model error increased from 35% to 55%. We also showed that the structured model capacity CDF followed the true CDF much more closely than did the Kronecker CDF.

The number of parameters for both models is a function of M_{Rx}, M_{Tx}, and D. We showed that, for all of these cases, the structured model requires fewer parameters than does the Kronecker model.

We then tested each model's ability to approximate the spatial structure of the channel. We plotted the MIMO APS for the structured and Kronecker models and compared them with the true MIMO APS. In most cases, we found that the structured model approximated the location and magnitude of scatterers better than did the Kronecker model. This was especially true in channels where the separability assumption did not hold, which was most often the case.

Finally, we discussed the significance of modeling the correlation between channels at different delays. The Kronecker model assumes complete independence between channels at different delays. By measuring the weight of the off-diagonal blocks of $\mathbf{R}_{WB,H}$, it was shown that the correlation across delay space was not significant for the WMSDR data but very significant for the BYU data. The Kronecker model error increased 12% from one data set to the next, whereas the structured model error remained relatively constant. This result implies that the increased correlation across delay space may be at least partly responsible for the increased Kronecker model error.

In the end, given two very different real-life data sets, and using the above metrics, the structured model consistently outperforms the Kronecker model.

6.7   NOTES AND REFERENCES

The MATLAB scripts used to produce the above analyses for the structured and Kronecker models, along with the WMSDR data set used in this chapter, is available online at http://soma.mcmaster.ca/~costa.

The people from the Wireless Communications Research lab at BYU have made part of their extensive narrowband and wideband database available online (125). Whereas some of the database is not yet public, the wideband BYU data used in this chapter is freely available.

APPENDIX A

AN INTRODUCTION TO TENSOR ALGEBRA

This appendix presents a brief introduction to tensor algebra and the *higher-order singular value decomposition* (HOSVD). Tensor algebra is used in Chapter 3 to extend the idea of correlation to receive-transmit-delay space and to develop a novel wideband MIMO channel model.

Multiple-antenna channels are often expressed using multidimensional quantities such as vectors, matrices, and tensors. Here, we include a brief overview of tensor algebra and introduce notations used throughout the text. This section focuses specifically on notations included in this text only and as a result is very limited in scope. It is not necessary to have a good understanding of tensor algebra to understand the subjects covered in this text. It is, however, important to understand the notation and operations covered in this section to do so. Tensor algebra is a very mature topic and is the subject of many textbooks and theses, including Refs. 136 and 137.

The chapter begins by discussing the HOSVD, which is a multidimensional extension to the matrix *singular value decomposition* (SVD). In presenting the HOSVD, we summarize some important tensor operations used in the rest of the text. The HOSVD also serves as the inspiration for the structured model.

A.1 NOTATION AND TENSOR ORDER

We denote scalars using lowercase letters with no indices attached; for example, $\{a, b, c, \ldots\}$. We denote vectors using bold lowercase letters $\{\mathbf{a}, \mathbf{b}, \mathbf{c}, \ldots\}$. We denote matrices using bold uppercase letters $\{\mathbf{A}, \mathbf{B}, \mathbf{C}, \ldots\}$.

Tensors are multidimensional generalizations of matrices. The *order* of a tensor is the number of dimensions needed to address a single element of a tensor; that is,

$$\mathcal{A} \in \mathbb{R}^{I_1 \times I_2 \times \cdots \times I_N} \tag{A.1}$$

Multiple-Input, Multiple-Output Channel Models: Theory and Practice. By Nelson Costa and Simon Haykin
Copyright © 2010 John Wiley & Sons, Inc.

is an Nth-order tensor. Scalars, vectors, and matrices are examples of zeroth-, first-, and second-order tensors, respectively. To avoid confusion, we always denote zeroth-, first-, and second-order tensors as scalars, vectors, and matrices, respectively. For the most part, higher-order tensors ($N > 2$) will be addressed using uppercase calligraphic letters $\{\mathcal{A}, \mathcal{B}, \mathcal{C}, \ldots\}$ except where it is otherwise noted.

The elements of a tensor of any order can be addressed by one of two ways; for example, the (i_1, i_2, i_3)th element of the tensor $\mathcal{A} \in \mathbb{C}^{I_1 \times I_2 \times I_3}$ is addressed as either $(\mathcal{A})_{i_1 i_2 i_3}$ or $a_{i_1 i_2 i_3}$ for indices $i_1 = \{1, \ldots, I_1\}$, $i_2 = \{1, \ldots, I_2\}$, and $i_3 = \{1, \ldots, I_3\}$. Throughout the text, we use a calligraphic uppercase letter (e.g., $\mathcal{A}$) or indexed lowercase letter (e.g., $a_{i_1 i_2 i_3}$) to refer to the same tensor interchangeably. The expression

$$\mathcal{A} \in \mathbb{C}^{I_1 \times \cdots \times I_n \times \cdots I_N} \tag{A.2}$$

denotes an Nth-order tensor.

A.2 THE HOSVD AND RELEVANT TENSOR ALGEBRA

The HOSVD is an extension of the two-dimensional (matrix) SVD (75). In the same way the matrix SVD provides an orthogonal basis for the row- and column-space of an arbitrary matrix, the HOSVD provides an orthogonal basis for each dimension of a higher-dimension tensor. It has been used, for example, in the analysis of higher-dimensional data in the computer vision problem (138).

A.2.1 Tensor Outer Product and Tensor Rank

The outer product of vectors $\mathbf{u}_1, \mathbf{u}_2, \ldots, \mathbf{u}_N$ is denoted by

$$\mathcal{B} = \mathbf{u}_1 \circ \mathbf{u}_2 \circ \cdots \circ \mathbf{u}_N. \tag{A.3}$$

The elements of $\mathcal{B}$ are found by multiplying each element in each vector as $b_{i_1 i_2 \cdots i_N} = u_{i_1} u_{i_2} \cdots u_{i_N}$. $\mathcal{B}$ is considered to be rank-1 because it can be expressed as a single term involving the outer product of N vectors. In general, analogous to the matrix case, the rank of a tensor is the minimum number of rank-1 tensors needed to express it as a linear combination. That is, the rank of $\mathcal{A}$ is R if we can express it as

$$\mathcal{A} = \sum_{r=1}^{R} \sigma_r (u_1^{(r)} \circ \cdots \circ u_N^{(r)}). \tag{A.4}$$

A.2.2 Matrix Unfolding

There are many ways to map a multidimensional tensor to a matrix. One set of such mappings are the *matrix unfolding* of $\mathcal{A}$. Specifically, the matrix unfolding

$$\mathbf{A}_{(n)} \in \mathbb{R}^{I_n \times (I_{n+1} \cdots I_N I_1 \cdots I_{n-1})} \tag{A.5}$$

is formed by stacking the columns formed by the nth dimension, one after the other to form a matrix. In the case of a third-order tensor, the process is illustrated in Fig. A.1. The n-rank of $\mathcal{A}$ is defined as the rank of the nth unfolding of $\mathcal{A}$; that is,

$$\begin{aligned} R_n &= \mathrm{rank}_n(\mathcal{A}) \\ &= \mathrm{rank}(\mathbf{A}_{(n)}). \end{aligned} \tag{A.6}$$

A.2.3   *N*-mode Product

Multiplication between a tensor and a matrix can also be defined. Consider the tensor $\mathcal{A} \in \mathbb{R}^{I_1 \times I_2 \times \cdots \times I_n \times \cdots \times I_N}$ and the matrix $\mathbf{M} \in \mathbb{R}^{J_n \times I_n}$. The *n-mode product* between $\mathcal{A}$ and $\mathbf{M}$ is defined as

$$\mathcal{B} = \mathcal{A} \times_n \mathbf{M}, \tag{A.7}$$

where the product tensor $\mathcal{B} \in \mathbb{R}^{I_1 \times \cdots \times I_{n-1} \times J_n \times I_{n+1} \times \cdots \times I_N}$. Element-wise, the n-mode product can be computed as

$$b_{i_1 \cdots i_{n-1} j_n i_{n+1} \cdots i_N} = \sum_{i_n} a_{i_1 \cdots i_n \cdots i_N} m_{j_n i_n}. \tag{A.8}$$

Using the nth unfolding of $\mathcal{A}$, we can also express the product as a matrix multiplication,

$$\mathbf{B}_{(n)} = \mathbf{M} \mathbf{A}_{(n)}. \tag{A.9}$$

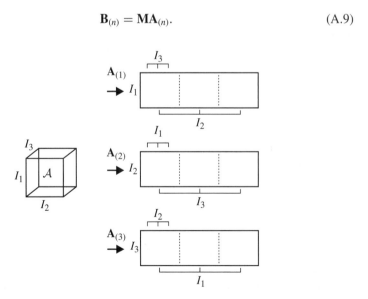

Figure A.1   Example matrix unfolding of a third-order tensor.

A.2.4   Tensor Summation Convention

The *tensor summation convention* provides a compact way of expressing a linear combination of tensor elements. It states that whenever an index appears twice in an expression, a summation over that index is implied. The summation is performed over the entire range of the repeated index.

As an example, consider the input vector $\mathbf{x} \in \mathbb{C}^2$, channel matrix $\mathbf{H} \in \mathbb{C}^{2 \times 2}$, and output vector $\mathbf{y} \in \mathbb{C}^2$. The output for each m can be expressed as the linear combination of input and channel tensor elements, that is,

$$
\begin{aligned}
y_1 &= h_{11}x_1 + h_{12}x_2 \\
y_2 &= h_{21}x_1 + h_{22}x_2.
\end{aligned}
\tag{A.10}
$$

Let m be a *free index* and n a *summation index*. The above system of equations can be rewritten as

$$
y_m = \sum_{n=1}^{2} h_{mn}x_n,
\tag{A.11}
$$

for $m = 1, 2$ and $n = 1, 2$. Using the summation convention, the summation symbol can be omitted, so that

$$
y_m = h_{mn}x_n.
\tag{A.12}
$$

A.2.5   Tensor Inner Product

Given two tensors $\mathcal{A} \in \mathbb{C}^{I_1 \times I_2 \times \cdots \times I_N}$ and $\mathcal{B} \in \mathbb{C}^{J_1 \times J_2 \times \cdots \times J_M}$, the inner product of both tensors over a common index $i_n = j_m = k$ is denoted as

$$
\mathcal{C} = \langle \mathcal{A}, \mathcal{B} \rangle_{n,m}.
\tag{A.13}
$$

Written explicitly,

$$
c_{i_1 \cdots i_{n-1} i_{n+1} \cdots i_N j_1 \cdots j_{m-1} j_{m+1} \cdots j_M} = \sum_{k} a_{i_1 \cdots i_{n-1} i_k i_{n+1} \cdots i_N} b_{j_1 \cdots j_{m-1} j_k j_{m+1} \cdots j_M},
\tag{A.14}
$$

where the resulting tensor $\mathcal{C} \in \mathbb{C}^{I_1 \times \cdots \times I_{n-1} \times I_{n+1} \times \cdots \times I_N \times J_1 \times \cdots \times J_{m-1} \times J_{m+1} \times \cdots \times J_M}$. Note that in order for the inner product to be defined, the size of the nth dimension of $\mathcal{A}$ must be equal to the size of the mth dimension of $\mathcal{B}$. Also, if the size of more than one (or all) dimension of both tensors match, it is possible to define the inner product over more than one dimension.

A.2.6 Tensor Scalar Product and the Frobenius Norm

Consider two tensors whose dimensions are equal, say $A, B \in \coprod^{I_1 \times I_2 \times \cdots \times I_N}$. The inner product over all dimensions results in a scalar value. That is, let the inner product of tensors A and B be

$$\langle A, B \rangle \triangleq \sum_{i_1} \cdots \sum_{i_N} a_{i_1 \cdots i_N} b^*_{i_1 \cdots i_N}. \tag{A.15}$$

Using this, the definition of a Frobenius norm can be extended to tensors. Specifically the Frobenius norm of $A \in \mathbb{C}^{I_1 \times I_2 \times \cdots \times I_N}$ is defined as

$$\|A\|_F \triangleq \sqrt{\langle A, A \rangle}. \tag{A.16}$$

Analogous to the matrix case, the square of the Frobenius norm is viewed as the "energy" of the tensor; that is, $\varepsilon_A = \|A\|_F^2$.

A.2.7 Computing the HOSVD

Using the concepts above, an Nth-order tensor can be decomposed using the HOSVD as,

$$A = S \times_1 \mathbf{U}_1 \times_2 \mathbf{U}_2 \cdots \times_N \mathbf{U}_N, \tag{A.17}$$

where $S \in \mathbb{R}^{I_1 \times I_2 \times \cdots \times I_N}$ is the *core tensor*, and $\mathbf{U}_1, \ldots, \mathbf{U}_N$ are N orthogonal spaces, one for each dimension of A. Each $\mathbf{U}_n$ forms an orthogonal basis for the column space of $\mathbf{A}_{(n)}$, and the elements in S describe the interaction between the bases. Unlike the matrix case, the core tensor is not "diagonal" in any sense; none of the elements in S are identically zero all the time. However, the core tensor does exhibit orthogonality.

Consider the subtensor $S_{i_n=k}$ formed by taking all the elements of S while keeping the dimension $i_n = k$. The subtensors follow the property

$$\langle S_{i_n=k}, S_{i_n=j} \rangle = 0, \quad \text{for } k \neq j. \tag{A.18}$$

Any tensor that follows this property for all $n, j, k \in \{1, \ldots, N\}$, is termed an *all-orthogonal* tensor. Also, the core tensor is *ordered* in that the Frobenius norm of the subtensors can be arranged according to

$$\|S_{i_n=1}\|_F \geq \|S_{i_n=2}\|_F \geq \cdots \geq \|S_{i_n=I_n}\|_F \geq 0 \tag{A.19}$$

for $n \in \{1, \ldots, N\}$.

The HOSVD of A is first computed by taking the matrix SVD of the nth unfolding of A as

$$\mathbf{A}_{(n)} = \mathbf{U}_n \mathbf{S}_n \mathbf{V}_n^T \tag{A.20}$$

for $n = \{1, \ldots, N\}$. We then solve for the core tensor $\mathcal{S}$ using

$$\mathcal{S} = \mathcal{A} \times_1 U_1^T \times_2 U_2^T \cdots \times_N U_N^T. \tag{A.21}$$

A.3   SUMMARY

Tensor calculus is a multidimensional generalization of matrix algebra. It is very useful in the analysis of multidimensional linear systems, such as the wideband MIMO channel. In particular, the HOSVD can be used to decompose a higher-order tensor into several orthogonal bases, one for each dimension of the tensor, and a core tensor that describes the interaction between the bases. The nth orthogonal basis is computed by computing the SVD of the tensor's nth unfolding. The HOSVD, and the relevant tensor algebra, provide the inspiration for the structured model.

APPENDIX B

PROOF OF THEOREMS FROM CHAPTER 3

This appendix presents proofs for key steps in the derivation of the structured model. It provides some insight into the mapping between tensor and matrix elements, the derivation of the wideband coupling coefficients, and the structured model synthesis equation.

B.1 PROOF OF (3.53)

We begin by hypothesizing that the Kronecker structure will allow us to line up the proper (m, n, d)th elements of the (i, j, k)th eigenvectors with the (m, n, d)th element of $\mathcal{H}$, and we write the multiplication out explicitly to make sure that we get to the summation formula:

$$(\mathbf{u}_{\text{Del},k} \otimes \mathbf{u}_{\text{Tx},j} \otimes \mathbf{u}_{\text{Rx},i})^{H} \text{vec}(\mathcal{H})$$

$$= \left[\begin{pmatrix} u_{\text{Del},1k} \\ \vdots \\ u_{\text{Del},Dk} \end{pmatrix} \otimes \begin{pmatrix} u_{\text{Tx},1j} \\ \vdots \\ u_{\text{Tx},M_{\text{Tx}}j} \end{pmatrix} \otimes \begin{pmatrix} u_{\text{Rx},1k} \\ \vdots \\ u_{\text{Rx},M_{\text{Rx}}k} \end{pmatrix} \right]^{H} \times \begin{pmatrix} h_{111} \\ \vdots \\ h_{M_{\text{Rx}}M_{\text{Tx}}D} \end{pmatrix}. \qquad \text{(B.1)}$$

Multiple-Input, Multiple-Output Channel Models: Theory and Practice. By Nelson Costa and Simon Haykin
Copyright © 2010 John Wiley & Sons, Inc.

Using the definition for the Kronecker product, we multiply out each element.

$$(\mathbf{u}_{\text{Del},k} \otimes \mathbf{u}_{\text{Tx},j} \otimes \mathbf{u}_{\text{Rx},i})^H \text{vec}(\mathcal{H})$$

$$
= \left(
\begin{pmatrix}
u_{\text{Del},1k} \cdot
\begin{pmatrix}
u_{\text{Tx},1j} \cdot \begin{pmatrix} u_{\text{Rx},1k} \\ \vdots \\ u_{\text{Rx},M_{\text{Rx}}k} \end{pmatrix} \\
\vdots \\
u_{\text{Tx},M_{\text{Tx}}j} \cdot \begin{pmatrix} u_{\text{Rx},1k} \\ \vdots \\ u_{\text{Rx},M_{\text{Rx}}k} \end{pmatrix}
\end{pmatrix} \\
\vdots \\
u_{\text{Del},Dk} \cdot
\begin{pmatrix}
u_{\text{Tx},1j} \cdot \begin{pmatrix} u_{\text{Rx},1k} \\ \vdots \\ u_{\text{Rx},M_{\text{Rx}}k} \end{pmatrix} \\
\vdots \\
u_{\text{Tx},M_{\text{Tx}}j} \cdot \begin{pmatrix} u_{\text{Rx},1k} \\ \vdots \\ u_{\text{Rx},M_{\text{Rx}}k} \end{pmatrix}
\end{pmatrix}
\end{pmatrix}
\right)^H
\times \begin{pmatrix} h_{111} \\ \vdots \\ h_{M_{\text{Rx}}M_{\text{Tx}}D} \end{pmatrix}
$$

(B.2)

$$
= \begin{pmatrix}
u_{\text{Rx},1k}u_{\text{Tx},1j}u_{\text{Del},1k} \\
\vdots \\
u_{\text{Rx},M_{\text{Rx}}k}u_{\text{Tx},1j}u_{\text{Del},1k} \\
u_{\text{Rx},1k}u_{\text{Tx},2j}u_{\text{Del},1k} \\
\vdots \\
u_{\text{Rx},M_{\text{Rx}}k}u_{\text{Tx},2j}u_{\text{Del},1k} \\
\vdots \\
u_{\text{Rx},M_{\text{Rx}}k}u_{\text{Tx},M_{\text{Tx}}j}u_{\text{Del},1k} \\
u_{\text{Rx},1k}u_{\text{Tx},1j}u_{\text{Del},2k} \\
\vdots \\
u_{\text{Rx},M_{\text{Rx}}k}u_{\text{Tx},M_{\text{Tx}}j}u_{\text{Del},Dk}
\end{pmatrix}^H
\times \begin{pmatrix} h_{111} \\ \vdots \\ h_{M_{\text{Rx}}M_{\text{Tx}}D} \end{pmatrix}
$$

$$
= \sum_{m=1}^{M_{\text{Rx}}} \sum_{n=1}^{M_{\text{Tx}}} \sum_{d=1}^{D} h_{mnd} u_{\text{Rx},mi}^* u_{\text{Tx},nj}^* u_{\text{Rx},dk}^* \cdot \qquad \blacksquare
$$

B.2 PROOF OF (3.54)

We start off by expanding the definition of the wideband coupling coefficient ω_{ijk},

$$\omega_{ijk} = \mathrm{E}\left\{\left(\sum_{m=1}^{M_{\mathrm{Rx}}}\sum_{n=1}^{M_{\mathrm{Tx}}}\sum_{d=1}^{D} h_{mnd} u_{\mathrm{Rx},mi}^* u_{\mathrm{Tx},nj}^* u_{\mathrm{Rx},dk}^*\right)^*\left(\sum_{p=1}^{M_{\mathrm{Rx}}}\sum_{q=1}^{M_{\mathrm{Tx}}}\sum_{r=1}^{D} h_{pqr} u_{\mathrm{Rx},pi}^* u_{\mathrm{Tx},qj}^* u_{\mathrm{Rx},rk}^*\right)\right\}$$

$$= \mathrm{E}\left\{\left[(\mathbf{u}_{\mathrm{Del},k}\otimes\mathbf{u}_{\mathrm{Tx},j}\otimes\mathbf{u}_{\mathrm{Rx},i})^H\mathrm{vec}(\mathcal{H})\right]^H\left[(\mathbf{u}_{\mathrm{Del},k}\otimes\mathbf{u}_{\mathrm{Tx},j}\otimes\mathbf{u}_{\mathrm{Rx},i})^H\mathrm{vec}(\mathcal{H})\right]\right\}$$

$$= \mathrm{E}\left\{\left[\mathrm{vec}^H(H)(\mathbf{u}_{\mathrm{Del},k}\otimes\mathbf{u}_{\mathrm{Tx},j}\otimes\mathbf{u}_{\mathrm{Rx},i})\right][(\mathbf{u}_{\mathrm{Del},k}\otimes\mathbf{u}_{\mathrm{Tx},j}\otimes\mathbf{u}_{\mathrm{Rx},i})^H\mathrm{vec}(\mathcal{H})]\right\}. \quad (B.3)$$

Because both terms in the square brackets are scalars, we can therefore switch their order.

$$\omega_{ijk} = \mathrm{E}\left\{\left[(\mathbf{u}_{\mathrm{Del},k}\otimes\mathbf{u}_{\mathrm{Tx},j}\otimes\mathbf{u}_{\mathrm{Rx},i})^H\mathrm{vec}(\mathcal{H})][\mathrm{vec}^H(\mathcal{H})(\mathbf{u}_{\mathrm{Del},k}\otimes\mathbf{u}_{\mathrm{Tx},j}\otimes\mathbf{u}_{\mathrm{Rx},i})]\right\}. \quad (B.4)$$

The elements of $\mathcal{H}$ are random variables, but the elements of the eigenvectors are not, therefore

$$\omega_{ijk} = (\mathbf{u}_{\mathrm{Del},k}\otimes\mathbf{u}_{\mathrm{Tx},j}\otimes\mathbf{u}_{\mathrm{Rx},i})^H\mathrm{E}\{\mathrm{vec}(\mathcal{H})\mathrm{vec}^H(\mathcal{H})\}(\mathbf{u}_{\mathrm{Del},k}\otimes\mathbf{u}_{\mathrm{Tx},j}\otimes\mathbf{u}_{\mathrm{Rx},i}) \quad (B.5)$$

$$= (\mathbf{u}_{\mathrm{Del},k}\otimes\mathbf{u}_{\mathrm{Tx},j}\otimes\mathbf{u}_{\mathrm{Rx},i})^H\mathbf{R}_{\mathrm{WB},H}(\mathbf{u}_{\mathrm{Del},k}\otimes\mathbf{u}_{\mathrm{Tx},j}\otimes\mathbf{u}_{\mathrm{Rx},i}) \qquad \blacksquare$$

B.3 PROOF OF (3.56)

We begin by computing the *eigenvalue decomposition* (EVD) of $\mathbf{R}_{\mathrm{WB,struct}}$ as

$$\mathbf{R}_{\mathrm{WB,struct}} = \mathbf{U}_{\mathrm{WB}}\cdot\mathrm{diag}(\omega_{111},\ldots,\omega_{M_{\mathrm{Rx}}M_{\mathrm{Tx}}D})\cdot\mathbf{U}_{\mathrm{WB}}^H$$

$$= \mathbf{U}_{\mathrm{WB}}\Omega_{\mathrm{WB}}\mathbf{U}_{\mathrm{WB}}^H. \quad (B.6)$$

Let Ω_{ij} denote the (i,j)th element of Ω_{WB}, and let $\mathbf{u}_{\mathrm{WB},i}$ be the ith column of $\mathbf{U}_{\mathrm{WB}}$. Also, let $\mathbf{A} = \mathbf{U}_{\mathrm{WB}}\Omega_{\mathrm{WB}}$, and let the columns of $\mathbf{A}$ be $\mathbf{a}_j = \sum_{i=1}^{M_{\mathrm{Rx}}M_{\mathrm{Tx}}D}\Omega_{ij}\mathbf{u}_{\mathrm{WB},i}$ so that

$$\mathbf{R}_{\mathrm{WB,struct}} = \mathbf{A}\mathbf{U}_{\mathrm{WB}}^H$$

$$= \sum_{j=1}^{M_{\mathrm{Rx}}M_{\mathrm{Tx}}D}\mathbf{a}_j\mathbf{u}_{\mathrm{WB},j}^H$$

$$= \sum_{i=1}^{M_{\mathrm{Rx}}M_{\mathrm{Tx}}D}\sum_{j=1}^{M_{\mathrm{Rx}}M_{\mathrm{Tx}}D}\omega_{ij}\mathbf{u}_{\mathrm{WB},i}\mathbf{u}_{\mathrm{WB},j}^H, \quad (B.7)$$

but we know that $\omega_{ij} = 0$ for $i \neq j$, so

$$\mathbf{R}_{\text{WB,struct}} = \sum_{i=1}^{M_{\text{Rx}}M_{\text{Tx}}D} \omega_{ii}\mathbf{u}_{\text{WB},i}\mathbf{u}_{\text{WB},i}^{H}. \tag{B.8}$$

We choose the columns $\mathbf{u}_{\text{WB},i}$ from the set of vectors resulting from the Kronecker product of the one-sided eigenvectors,

$$\mathbf{u}_{\text{WB},i} \in \left\{ \underbrace{(\mathbf{u}_{\text{Del},1} \otimes \mathbf{u}_{\text{Tx},1} \otimes \mathbf{u}_{\text{Rx},1})}_{i=1}, \right.$$
$$\left. \underbrace{(\mathbf{u}_{\text{Del},1} \otimes \mathbf{u}_{\text{Tx},1} \otimes \mathbf{u}_{\text{Rx},2})}_{i=2}, \ldots, \underbrace{(\mathbf{u}_{\text{Del},D} \otimes \mathbf{u}_{\text{Tx},M_{\text{Tx}}} \otimes \mathbf{u}_{\text{Rx},M_{\text{Rx}}})}_{i=M_{\text{Rx}}M_{\text{Tx}}D} \right\}. \tag{B.9}$$

The result is that

$$\mathbf{R}_{\text{WB,struct}} = \sum_{i=1}^{M_{\text{Rx}}} \sum_{j=1}^{M_{\text{Tx}}} \sum_{k=1}^{D} \omega_{ijk}(\mathbf{u}_{\text{Del},k} \otimes \mathbf{u}_{\text{Tx},j} \otimes \mathbf{u}_{\text{Rx},i})(\mathbf{u}_{\text{Del},k} \otimes \mathbf{u}_{\text{Tx},j} \otimes \mathbf{u}_{\text{Rx},i})^{H}. \tag{B.10}$$

Note that this is essentially the same proof as for the EVD (i.e., $\mathbf{A} = \mathbf{U}\boldsymbol{\Lambda}\mathbf{U}^{H} = \sum_i \lambda_i \mathbf{u}_i \mathbf{u}_i^{H}$) except that we are mapping elements from three dimensions to form a matrix. ■

B.4 PROOF OF (3.57)

We first look at the narrowband case as given by the Weichselberger model, extend the synthesis equation to the wideband case, and show how the extension leads to (3.57). Then we check that

$$E\{\text{vec}(\mathcal{H}_{\text{struct}})\text{vec}^{H}(\mathcal{H}_{\text{struct}})\} = \mathbf{R}_{\text{WB,struct}}, \tag{B.11}$$

which completes the proof. We know that, in the case of a narrowband MIMO channel, we can synthesize each element of the H-matrix using

$$h_{mn} = \sum_{i=1}^{M_{\text{Rx}}} \sum_{j=1}^{M_{\text{Tx}}} w_{ij} u_{\text{Rx},mi} u_{\text{Tx},nj}, \tag{B.12}$$

where $w_{ij} = g_{ij}\sqrt{\omega_{ij}}$, g_{ij} is a complex-Gaussian random variable, and ω_{ij} is the narrowband coupling coefficient, as defined in Ref. 51. We propose that we can extend this definition to include a third dimension, so that each element h_{mnd} can be expressed as

$$h_{mnd} = \sum_{i=1}^{M_{\text{Rx}}} \sum_{j=1}^{M_{\text{Tx}}} \sum_{k=1}^{D} w_{ijk} u_{\text{Rx},mi} u_{\text{Tx},nj} u_{\text{Del},dk}, \tag{B.13}$$

where $w_{ijk} = g_{ijk}\sqrt{\omega_{ijk}}$, g_{ijk} are complex-Gaussian random variables, and ω_{ijk} are the wideband coupling coefficients.

Consider the following identity. If $\mathcal{B} = \mathcal{A} \times_n \mathbf{M}$, then we know that each element of $\mathcal{B}$ can be found using $(\mathcal{B})_{i_1...i_{n-1}j_n i_{n+1}...i_N} = \sum_{i_n} a_{i_1...i_{n-1}i_n i_{n+1}...i_N} m_{j_n i_n}$, for tensors $\mathcal{A} \in \mathbb{C}^{I_1 \times \cdots \times I_n \times \cdots \times I_N}$, $\mathcal{B} \in \mathbb{C}^{I_1 \times \cdots \times J_n \times \cdots \times I_N}$, and matrix $\mathbf{M} \in \mathbb{C}^{J_n \times I_n}$. Let $\mathcal{A} = \mathcal{G} \times_1 \mathbf{U}_{Rx}$ so that

$$a_{mjk} = \sum_{i=1}^{M_{Rx}} g_{ijk} u_{Rx,mi}. \tag{B.14}$$

Similarly, let $\mathcal{B} = \mathcal{A} \times_2 \mathbf{U}_{Tx}$ so that

$$b_{mnk} = \sum_{j=1}^{M_{Tx}} a_{mjk} u_{Tx,nj}$$
$$= \sum_{i=1}^{M_{Rx}} \sum_{j=1}^{M_{Tx}} w_{ijk} u_{Rx,mi} u_{Tx,nj}. \tag{B.15}$$

Finally, let $\mathcal{C} = \mathcal{B} \times_3 \mathbf{U}_{Del}$ so that

$$c_{mnd} = \sum_{j=1}^{M_{Tx}} b_{mnk} u_{Del,dk}$$
$$= \sum_{i=1}^{M_{Rx}} \sum_{j=1}^{M_{Tx}} \sum_{k=1}^{D} w_{ijk} u_{Rx,mi} u_{Tx,nj} u_{Del,dk} \tag{B.16}$$
$$= h_{mnd}.$$

Therefore, we have shown that

$$\mathcal{H}_{struct} = \mathcal{G} \times_1 \mathbf{U}_{Rx} \times_2 \mathbf{U}_{Tx} \times_3 \mathbf{U}_{Del}$$
$$\Rightarrow h_{mnd} = \sum_{i=1}^{M_{Rx}} \sum_{j=1}^{M_{Tx}} \sum_{k=1}^{D} w_{ijk} u_{Rx,mi} u_{Tx,nj} u_{Del,dk}. \tag{B.17}$$

Now we need to show that $\mathrm{E}\{\mathrm{vec}(\mathcal{H}_{struct})\mathrm{vec}^H(\mathcal{H}_{struct})\} = \mathbf{R}_{WB,struct}$. First we define a correlation tensor $\mathcal{R} \in \mathbb{C}^{M_{Rx} \times M_{Tx} \times D \times M_{Rx} \times M_{Tx} \times D}$ such that

$$r_{mndpqr} = \mathrm{E}\{h_{mnd} h_{pqr}^*\}$$
$$= \mathrm{E}\left\{\left(\sum_{ijk} w_{ijk} u_{Rx,mi} u_{Tx,nj} u_{Del,dk}\right)\left(\sum_{stu} w_{stu} u_{Rx,ps} u_{Tx,qt} u_{Del,ru}\right)^*\right\} \tag{B.18}$$
$$= \sum_{ijkstu} \mathrm{E}\{w_{ijk} w_{stu}^*\}(u_{Rx,mi} u_{Tx,nj} u_{Del,dk})(u_{Rx,ps} u_{Tx,qt} u_{Del,ru})^*.$$

We know that

$$
\begin{aligned}
E\{w_{ijk}w_{stu}^*\} &= E\left\{\left(g_{ijk}\sqrt{\omega_{ijk}}\right)\left(g_{stu}\sqrt{\omega_{stu}}\right)^*\right\} \\
&= E\{g_{ijk}g_{stu}^*\}\sqrt{\omega_{ijk}}\sqrt{\omega_{stu}}.
\end{aligned}
\tag{B.19}
$$

By definition,

$$
E\{g_{ijk}g_{stu}^*\} = \begin{cases} 0 & \text{for } i\neq s \text{ and } j\neq t \text{ and } k\neq u \\ 1 & \text{for } i=s,\, j=t, \text{ and } k=u \end{cases},
\tag{B.20}
$$

so that

$$
r_{mndpqr} = \sum_{ijk}\omega_{ijk}(u_{\mathrm{Rx},mi}u_{\mathrm{Tx},nj}u_{\mathrm{Del},dk})(u_{\mathrm{Rx},pi}u_{\mathrm{Tx},qj}u_{\mathrm{Del},rk})^*.
\tag{B.21}
$$

There are many ways to map (B.21) to a matrix. One such mapping results in the equation for the wideband correlation matrix. We can map the elements r_{mndpqr} such that

$$
\mathbf{R}_{\mathrm{WB,struct}} = \left(
\begin{array}{cccccccc}
r_{111111} & \cdots & r_{111M_{\mathrm{Rx}}11} & r_{111121} & \cdots \\
\vdots & & \vdots & \vdots & \\
r_{M_{\mathrm{Rx}}11111} & \cdots & r_{M_{\mathrm{Rx}}11M_{\mathrm{Rx}}11} & r_{M_{\mathrm{Rx}}11121} & \cdots \\
r_{121111} & \cdots & r_{121M_{\mathrm{Rx}}11} & r_{121121} & \cdots \\
\vdots & & \vdots & \vdots & \\
r_{M_{\mathrm{Rx}}M_{\mathrm{Tx}}1111} & \cdots & r_{M_{\mathrm{Rx}}M_{\mathrm{Tx}}1M_{\mathrm{Rx}}11} & r_{M_{\mathrm{Rx}}M_{\mathrm{Tx}}1121} & \cdots \\
r_{112111} & \cdots & r_{112M_{\mathrm{Rx}}11} & r_{112121} & \cdots \\
\vdots & & \vdots & \vdots & \\
r_{M_{\mathrm{Rx}}M_{\mathrm{Tx}}D111} & \cdots & r_{M_{\mathrm{Rx}}M_{\mathrm{Tx}}DM_{\mathrm{Rx}}11} & r_{M_{\mathrm{Rx}}M_{\mathrm{Tx}}D121} & \cdots
\end{array}
\right.
$$

$$
\left.
\begin{array}{cccccccc}
r_{111M_{\mathrm{Rx}}M_{\mathrm{Tx}}1} & r_{111112} & \cdots & r_{111M_{\mathrm{Rx}}M_{\mathrm{Tx}}D} \\
\vdots & \vdots & & \vdots \\
r_{M_{\mathrm{Rx}}11M_{\mathrm{Rx}}M_{\mathrm{Tx}}1} & r_{M_{\mathrm{Rx}}11112} & \cdots & r_{M_{\mathrm{Rx}}11M_{\mathrm{Rx}}M_{\mathrm{Tx}}D} \\
r_{121M_{\mathrm{Rx}}M_{\mathrm{Tx}}1} & r_{121112} & \cdots & r_{121M_{\mathrm{Rx}}M_{\mathrm{Tx}}D} \\
\vdots & \vdots & & \vdots \\
r_{M_{\mathrm{Rx}}M_{\mathrm{Tx}}1M_{\mathrm{Rx}}M_{\mathrm{Tx}}1} & r_{M_{\mathrm{Rx}}M_{\mathrm{Tx}}1112} & \cdots & r_{M_{\mathrm{Rx}}M_{\mathrm{Tx}}1M_{\mathrm{Rx}}M_{\mathrm{Tx}}D} \\
r_{112M_{\mathrm{Rx}}M_{\mathrm{Tx}}1} & r_{112112} & \cdots & r_{112M_{\mathrm{Rx}}M_{\mathrm{Tx}}D} \\
\vdots & \vdots & & \vdots \\
r_{M_{\mathrm{Rx}}M_{\mathrm{Tx}}DM_{\mathrm{Rx}}M_{\mathrm{Tx}}1} & r_{M_{\mathrm{Rx}}M_{\mathrm{Tx}}D112} & \cdots & r_{M_{\mathrm{Rx}}M_{\mathrm{Tx}}DM_{\mathrm{Rx}}M_{\mathrm{Tx}}D}
\end{array}
\right).
\tag{B.22}
$$

Given that $r_{mndpqr} = \mathrm{E}\{h_{mnd}h_{pqr}^*\}$, we have

$$
\begin{aligned}
\mathbf{R}_{\mathrm{WB,struct}} &= \begin{pmatrix} \mathrm{E}\{h_{111}h_{111}^*\} & \cdots & \mathrm{E}\{h_{111}h_{M_{\mathrm{Rx}}M_{\mathrm{Tx}}D}^*\} \\ \vdots & \ddots & \vdots \\ \mathrm{E}\{h_{M_{\mathrm{Rx}}M_{\mathrm{Tx}}D}h_{111}^*\} & \cdots & \mathrm{E}\{h_{M_{\mathrm{Rx}}M_{\mathrm{Tx}}D}h_{M_{\mathrm{Rx}}M_{\mathrm{Tx}}D}^*\} \end{pmatrix} \\[2mm]
&= \mathrm{E}\left\{ \begin{pmatrix} h_{111}h_{111}^* & \cdots & h_{111}h_{M_{\mathrm{Rx}}M_{\mathrm{Tx}}D}^* \\ \vdots & \ddots & \vdots \\ h_{M_{\mathrm{Rx}}M_{\mathrm{Tx}}D}h_{111}^* & \cdots & h_{M_{\mathrm{Rx}}M_{\mathrm{Tx}}D}h_{M_{\mathrm{Rx}}M_{\mathrm{Tx}}D}^* \end{pmatrix} \right\} \\[2mm]
&= \mathrm{E}\{\mathrm{vec}(\mathcal{H})\mathrm{vec}^H(\mathcal{H})\}.
\end{aligned}
\tag{B.23}
$$

Given this structure, we can rewrite (B.21),

$$
\mathbf{R}_{WB,struct} = \sum_{i=1}^{M_{\mathrm{Rx}}}\sum_{j=1}^{M_{\mathrm{Tx}}}\sum_{k=1}^{D} \omega_{ijk}(\mathbf{u}_{Del,k}\otimes\mathbf{u}_{Tx,j}\otimes\mathbf{u}_{Rx,i})(\mathbf{u}_{Del,k}\otimes\mathbf{u}_{Tx,j}\otimes\mathbf{u}_{Rx,i})^H.
\tag{B.24}
$$

Thus we have proved that $\mathrm{E}\{\mathrm{vec}(\mathcal{H}_{\mathrm{struct}})\mathrm{vec}^H(\mathcal{H}_{\mathrm{struct}})\} = \mathbf{R}_{\mathrm{WB,struct}}$, indicating that the synthesis equation is correct. ∎

APPENDIX C
COST 273 MODEL SUMMARY

C.1 COST 273 MODEL PARAMETERS

The model parameters fall under three categories: external, stochastic, and computed parameters. External parameters include fixed parameters, such as base station height, receiver location, and so forth. Stochastic parameters are used for random variables, for example, the mean and standard deviation of the cluster AoA. Computed parameters are those computed over the course of simulation. Both the external and stochastic parameters are defined in the COST 273 standard for each environment.

Symbol	Description	Reference
f_c	Carrier frequency	External
h_{Rx}	Receiver, transmitter station height	External
h_{Tx}		
$\mathbf{r}_{Rx} = \begin{bmatrix} x_{Rx} & y_{Rx} & h_{Rx} \end{bmatrix}$	Position vector	External
$\mathbf{r}_{Tx} = \begin{bmatrix} x_{Tx} & y_{Tx} & h_{Tx} \end{bmatrix}$		
$\mathbf{v}_{Rx}, \mathbf{v}_{Tx}$	Rx, Tx velocity	External
$G_m(\varphi, \theta), G_n(\varphi, \theta)$	Antenna gains, mth receive, nth transmit	External
$d_{Rx,Tx}, \varphi_{Rx,Tx}, \theta_{Rx,Tx}$	LoS path distance, azimuth, elevation	Computed
τ_0	LoS delay	Computed
$N_{c.local}$	Number of local clusters	Stochastic
$N_{c,add}$	Average number of additional clusters	Stochastic

(Continued)

Multiple-Input, Multiple-Output Channel Models: Theory and Practice. By Nelson Costa and Simon Haykin
Copyright © 2010 John Wiley & Sons, Inc.

Symbol	Description	Reference
N_c, $N_{c,\text{local}}$, $N_{c,\text{twin}}$, $N_{c,\text{single}}$, $N_{c,\text{add}}$	Number of clusters, local clusters, twin clusters, single-interaction clusters, and additional clusters	Computed
K_{sel}	Ratio of single-interacting to twin clusters	Stochastic
P_c	Power of the cth cluster	Computed
k_τ	Cluster attenuation coefficient	Stochastic
τ_c	Cluster delay	Computed
τ_b	Cutoff delay	Stochastic
S_c	Fading factor for cluster c	Computed
σ_s	Fading factor standard deviation	Stochastic
$\sigma_{\tau,c}$, $\sigma_{\varphi_{\text{Rx}},c}$, $\sigma_{\varphi_{\text{Tx}},c}$, $\sigma_{\theta_{\text{Rx}},c}$, $\sigma_{\theta_{\text{Tx}},c}$	Cluster spreads: delay, azimuth AoA, azimuth AoD, elevation AoA, elevation AoD	Computed
μ_τ, σ_τ	Delay spread mean, standard deviation	Stochastic
$\mu_{\varphi,\text{Rx}}$, $\mu_{\varphi,\text{Tx}}$, $\sigma_{\varphi,\text{Rx}}$, $\sigma_{\varphi,\text{Tx}}$	Azimuth spread mean, standard deviation	Stochastic
$\mu_{\theta,\text{Rx}}$, $\mu_{\theta,\text{Tx}}$, $\sigma_{\theta,\text{Rx}}$, $\sigma_{\theta,\text{Tx}}$	Elevation spread mean, standard deviation	Stochastic
$\rho_{x,y}$	Correlation coefficients for $x, y \in \{S_c,$ $\sigma_{\tau,c}, \sigma_{\varphi_{\text{Rx}},c}, \sigma_{\varphi_{\text{Tx}},c}, \sigma_{\theta_{\text{Rx}},c}, \sigma_{\theta_{\text{Tx}},c}\}$	Stochastic
$\varphi_{\text{Rx},c}$, $\theta_{\text{Rx},c}$, $\varphi_{\text{Tx},c}$, $\theta_{\text{Tx},c}$	Cluster azimuth and elevation AoA, azimuth and elevation AoD	Computed
$d_{\tau,c}$	Cluster diameter, delay domain	Computed
$\tau_{\text{Rx},c}$, $\tau_{\text{Tx},c}$, $\tau_{\text{link},c}$	Delay from the cluster to the receiver, from the transmitter to the cluster, and link delay (for twin clusters)	Computed
$a_{X,c}$, $b_{X,c}$, $h_{X,c}$	Cluster span in the delay, azimuth, and elevation domains	Computed
$\mathbf{r}_{\text{Rx},c} = \begin{bmatrix} x_{\text{Rx},c} & y_{\text{Rx},c} & h_{\text{Rx},c} \end{bmatrix}$ $\mathbf{r}_{\text{Tx},c} = \begin{bmatrix} x_{\text{Tx},c} & y_{\text{Tx},c} & h_{\text{Tx},c} \end{bmatrix}$	Cluster Cartesian position from the receiver's, transmitter's point of view	Computed
$d_{\text{Rx},c}$ $d_{\text{Tx},c}$	Cluster distance from the receiver, transmitter	Computed
$r_{\min}$	Minimum cluster distance	Stochastic
σ_r	Cluster distance standard deviation	Stochastic
$\sigma_{\varphi_{\text{Tx}},\text{single},c}$	Azimuth spread for single-interaction clusters	Stochastic
N_{MPC}	Number of multipath components per cluster (fixed)	Stochastic
$\mathbf{r}_{\text{Rx},c,\ell} = \begin{bmatrix} x_{\text{Rx},c,\ell} & y_{\text{Rx},c,\ell} & z_{\text{Rx},c,\ell} \end{bmatrix}$ $\mathbf{r}_{\text{Tx},c,\ell} = \begin{bmatrix} x_{\text{Tx},c,\ell} & y_{\text{Tx},c,\ell} & z_{\text{Tx},c,\ell} \end{bmatrix}$	Multipath component vectors, Rx, Tx, Cartesian coordinates	Computed
$\tau_{c,\ell}$	Multipath component delay	Computed
$\begin{bmatrix} \varphi_{\text{Rx},c,\ell} & \theta_{\text{Rx},c,\ell} & d_{\text{Rx},c,\ell} \end{bmatrix}$ $\begin{bmatrix} \varphi_{\text{Tx},c,\ell} & \theta_{\text{Tx},c,\ell} & d_{\text{Tx},c,\ell} \end{bmatrix}$	Multipath component vectors, Rx, Tx, polar coordinates	Computed
$P_{c,\ell}$	Multipath component power	Computed

(Continued)

Symbol	Description	Reference
K_{MPC}	Multipath component Ricean K-factor	Stochastic
μ_K, σ_K	Mean, standard deviation w.r.t. multipath component Ricean distribution	Stochastic
A_K	Amplitude of the LoS component w.r.t. multipath component Ricean distribution	Computed
A_{LoS}	LoS attenuation, multipath components	Computed
$A_{\text{Att},c,\ell}$	Multipath component attenuation	Computed
h_{LoS}, $h_{\text{NLoS},c,\ell}$	Multipath component LoS and NLoS contribution to impulse response	Computed
$h_{\text{LoS},m,n}$, $h_{\text{NLoS},m,n}$, $h_{m,n}$	LoS, NLoS, and overall channel impulse response at receiver m, transmitter n	Computed
R_c, L_c	Visibility region radius, transition region radius	Stochastic
$\mathbf{r}_c$	Visibility region center	Computed
L_X	Autocorrelation distance for $X \in \{S, \tau, \varphi_{\text{Rx}}, \varphi_{\text{Tx}}, \theta_{\text{Rx}}, \theta_{\text{Tx}}\}$	Stochastic
$\mu_{c,v}$, $\sigma_{c,v}$	Cluster movement mean, standard deviation	Stochastic

C.2 COST 273 ENVIRONMENTS

Environment	Propagation Scenario	Distinguishing Features
Macrocells: transmitter above rooftop	Small macrocells in urban center	Dense urban, tall buildings
	Large urban macrocells	Same as small macrocell, but transmitter antenna located farther above rooftop for greater coverage
	Outdoor-to-indoor urban	Similar to urban macrocells. Receiver located indoors, can be much higher than street level.
Microcells: transmitter height at or below rooftop	Urban center	Similar to macrocell. urban with lower transmitter height
Picocells: enclosed environments	Halls	Large enclosed space, $>100 \text{ m}^2$
	Office, LoS	Series of single offices along corridor, or hall with cubicles. Receiver, transmitter have LoS.
	Office, NLoS	Same as above, with NLoS

C.3 RCM MODEL SUMMARY

Symbol	Description
Stochastic parameters, Θ_{env}	
N_c	Number of clusters
$N_{c,\ell}$	Number of multipath components in cluster c
$\sigma_{\varphi_{\text{Rx}},c}$, $\sigma_{\varphi_{\text{Tx}},c}$, $\sigma_{\theta_{\text{Rx}},c}$, $\sigma_{\theta_{\text{Tx}},c}$, $\sigma_{\tau,c}$	Cluster spreads, azimuth, elevation, delay
$\sigma_{\gamma,c}^2$	Cluster power
ρ_c	Total snapshot power, all clusters
$\overline{\tau}_c$	Mean cluster delay
$\overline{\varphi}_{\text{Rx},c}$, $\overline{\varphi}_{\text{Tx},c}$	Mean cluster azimuth angles
$\overline{\theta}_{\text{Rx},c}$, $\overline{\theta}_{\text{Tx},c}$	Mean cluster elevation angles
$\Delta\overline{\varphi}_{\text{Rx},c}$, $\Delta\overline{\varphi}_{\text{Tx},c}$, $\Delta\overline{\theta}_{\text{Rx},c}$, $\Delta\overline{\theta}_{\text{Tx},c}$, $\Delta\overline{\tau}_c$, $\Delta\sigma_{\gamma,c}^2$	Rate of change parameters for mean AoA, AoD, delay, and power for cluster c
Λ_c	Cluster lifetime
Multipath component parameters, $\Theta_{c,\ell}$	
$\varphi_{\text{Rx},c,\ell}$, $\varphi_{\text{Tx},c,\ell}$	Multipath component azimuth angles
$\theta_{\text{Rx},c,\ell}$, $\theta_{\text{Tx},c,\ell}$	Multipath component elevation angles
$\tau_{c,\ell}$	Multipath component delay
$\gamma_{c,\ell}$	Complex multipath component gain
Additional cluster parameters, Θ_c	
$p(x_{\text{birth}})$	PDF, cluster births per Λ_c
Additional time parameters	
Δt_Λ	Cluster birth/death interval
Δt_s	Snapshot period
$\lvert\gamma_{\text{att}}\rvert^2$	Cluster attenuation parameter
External parameters	
$\mathbf{a}_{\text{Rx}}(\varphi_{\text{Rx}}, \theta_{\text{Rx}})$, $\mathbf{a}_{\text{Tx}}(\varphi_{\text{Tx}}, \theta_{\text{Tx}})$	Array steering vectors
B	System bandwidth
$\tau_{\max}$	Maximum resolvable delay
Δf	Frequency resolution
d_λ	Transmitter distance traveled for one snapshot
v_{Tx}	Transmitter velocity
$N_{\Delta f}$	Number frequency bins

C.4 FUNCTION ATAN2

Often, when converting from Cartesian to polar coordinates, we need to be able to map coordinates in all four quadrants to polar coordinates in some repeatable fashion. The atan2 function is a simple extension to the atan function and often implemented in programming languages such as MATLAB (139).

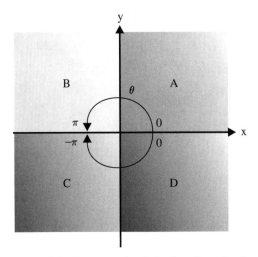

Figure C.1   Four quadrants for function atan2.

Table C.1   Summary for the function atan2($\cdot$)

Quadrant	Sign x, y	θ	Range: θ	Range: y/x
A	$x =$ positive $y =$ positive	$\theta = \text{atan}(y/x)$	$\left[0, \dfrac{\pi}{2}\right)$	$[0, +\infty)$
B	$x =$ negative $y =$ positive	$\theta = \text{atan}(y/x) + \pi$	$\left(\dfrac{\pi}{2}, \pi\right]$	$(-\infty, 0]$
C	$x =$ negative $y =$ negative	$\theta = \text{atan}(y/x) - \pi$	$\left(-\dfrac{\pi}{2}, -\pi\right]$	$(+\infty, 0]$
D	$x =$ positive $y =$ negative	$\theta = \text{atan}(y/x)$	$\left[0, -\dfrac{\pi}{2}\right)$	$[0, -\infty)$

Consider the two-dimensional case shown in Fig. C.1. The angle $\theta \in [-\pi, \pi]$ depends on the quadrant to which the point (x, y) belongs. The quadrant depends on the sign of x and y. Table C.1 summarizes how we compute atan2(y/x).

GLOSSARY

OPERATORS

$\otimes$	Kronecker product
$\odot$	Hadamard (element-wise) product
$\cdot^{*}$	Complex conjugate
$\cdot^{T}$	Matrix/vector transpose
$\cdot^{H}$	Matrix/vector Hermitian (complex) transpose
$\mathrm{vec}(\cdot)$	Mapping matrix/tensor elements to a vector
$\mathrm{unvec}(\cdot)$	Reverse mapping from vector to matrix/tensor
$\langle \cdot, \cdot \rangle_{n,m}$	Tensor inner product, index n of first tensor with index m of second
$\cdot \circ \cdot$	Tensor outer product
$\cdot \times_{n} \cdot$	n-mode product
$\mathbf{A}_{(n)}$	nth matrix unfolding
$\mathrm{E}\{\cdot\}$	Expectation operator
$\lvert \cdot \rvert$	Absolute value
$\lVert \cdot \rVert$	Vector norm
$\lVert \cdot \rVert_{2}$	Matrix 2-norm
$\lVert \cdot \rVert_{F}$	Matrix/tensor Frobenius norm
$\mathrm{rank}_{n}(\cdot)$	Tensor n-rank
$\cdot \leftrightharpoons \cdot$	Two continuous functions related via the Fourier transform
$\cdot \underset{\mathrm{DFT}}{\leftrightharpoons} \cdot$	Two discrete functions related via the discrete Fourier transform
$\cdot \underset{\mathrm{FFT}}{\leftrightharpoons} \cdot$	Two discrete functions related via the fast Fourier transform
$\mathrm{F}^{-1}\{\cdot\}$	Inverse Fourier transform

Multiple-Input, Multiple-Output Channel Models: Theory and Practice. By Nelson Costa and Simon Haykin
Copyright © 2010 John Wiley & Sons, Inc.

| · ∗ · | Convolution |
| $\mathbf{A}^{\frac{1}{2}}$ | Square-root matrix |

ABBREVIATIONS

3GPP	3rd Generation Partnership Project
ADC	Analog-to-digital converter
AGC	Automatic gain control
AoA	Angle of arrival
AoD	Angle of departure
APS	Azimuth power spectrum
AWGN	Additive white gaussian noise
BLAST	Bell Labs layered space–time
BPSK	Binary phase-shift keying
BS	Base station
BYU	Brigham Young University
CDF	Cumulative distribution function
CDMA	Code division multiple access
COST	European Cooperation in the field of Scientific and Technical Research
CSI	Channel state information
DAC	Digital-to-analog converter
DAQ	Data acquisition
DDS	Direct digital synthesizer
DFT	Discrete Fourier transform
DIO	Digital input–output
DNL	Differential nonlinearity error
DPWR	Digital power
DSP	Digital signal processing
ECM	Environmental characterization metric
EM	Electromagnetic
ENOB	Effective number-of-bits
EVD	Eigenvalue decomposition
FDTD	Finite-difference time domain
FFT	Fast Fourier transform
FIFO	First-in, first-out
FTW	Forschungszentrum Telekommunikation Wien
GCM	Generic channel model
GPS	Global Positioning System
GSM	Global system for mobile communications
HOSVD	Higher-order singular value decomposition
IF	Intermediate frequency
iFFT	Inverse fast Fourier transform
IID	Independent and identically distributed
IQ	In-phase/quadrature
ISI	Intersymbol interference
ISIS	Initialization and searching improved space-alternating generalized expectation maximization algorithm
KDE	Kernel density estimator

LED	Light emitting diode
LNA	Low-noise amplifier
LO	Local oscillator
LoS	Line-of-sight
MAP	Maximum a-posteriori
MCD	Multipath component distance
MCU	Microprocessor control unit
MIMO	Multiple-input, multiple-output
MISO	Multiple-input, single-output
ML	Maximum likelihood
ML sequence	Maximum-length sequence
MPC	Multipath component
M-PSK	M-ary phase-shift keying
M-QAM	M-ary quadrature amplitude modulation
MS	Mobile station
NLoS	Non–line-of-sight
PA	Power amplifier
PC	Personal computer
PCB	Printed circuit board
PDF	Probability distribution function
PDP	Power delay profile
PN sequence	Pseudo-noise sequence
QPSK	Quadrature phase-shift keying
RCM	Random cluster model
RF	Radio frequency
RFPWR	RF power
RMS	Root mean-square
RSSI	Root-square strength indication
Rx	Receiver
SAGE	Space-alternating generalized expectation maximization
SCM	Spatial channel model
SDR	Software defined radio
SIMO	Single-input, multiple-output
SISO	Single-input, single-output
SNR	Signal-to-noise ratio
SVD	Singular value decomposition
Tx	Transmitter
ULA	Uniform linear array
UWB	Ultra wide band
V-BLAST	Vertical Bell Labs layered space–time
VGA	Variable gain amplifier
VSNR	Vector space narrowband radiometer
WINNER	The Wireless World Initiative New Radio channel model
WMSDR	Wideband multiple-input, multiple-output software defined radio
WSS	Wide-sense stationary

BIBLIOGRAPHY

1. IEEE Virtual Museum. "Electromagnetism, Maxwell's Equations, and Microwaves." Available at: www.ieee-virtual-museum.org/exhibit/exhibit.php?taid=&id=159265 &id=1&seq=3&view=.

2. J. C. Maxwell. "On physical lines of force." *London Edinburgh Dublin Philos. Mag. J. Sci.*, March 1861.

3. Wikipedia. "Maxwell's Equations." Available at: http://en.wikipedia.org/wiki/Maxwell%27s_equations.

4. J. C. Maxwell. "A dynamical theory of the electromagnetic field." *Proc. R. Soc. London*, vol. 13, pp. 531–536, 1863.

5. J. C. Maxwell. *A Treatise on Electricity and Magnetism*, vol. I. London: Oxford University Press, 1873.

6. Wikipedia. "Heinrich Hertz." Available at: http://en.wikipedia.org/wiki/Heinrich_Hertz.

7. Wikipedia. "Invention of Radio." Available at: http://en.wikipedia.org/wiki/Invention_of_radio.

8. Wikipedia. "Nikola Tesla." Available at: http://en.wikipeida.org/wiki/Nikola_Tesla.

9. R. Naughton. "Adventures in Cybersound: Oliver Joseph Lodge, Sr: 1851–1940." Available at: http://www.acmi.net.au/AIC/LODGE_BIO.html.

10. B. S. Belrose. "Fessenden and Marconi: their differing technologies and the transatlantic experiments during the first decade of this century." http://www.ieee.ca/millennium/radio/radio_differences.html, 1995.

11. Wikipedia. "Guglielmo Marconi." Available at: http://en.wikipedia.org/wiki/Guglielmo_Marconi.

Multiple-Input, Multiple-Output Channel Models: Theory and Practice. By Nelson Costa and Simon Haykin
Copyright © 2010 John Wiley & Sons, Inc.

12. W. C. Jakes. *Mobile Communications*. New York: John Wiley & Sons, 1974.

13. J. H. Winters. "Smart antennas for wireless systems." *IEEE Personal Commun. Mag.*, vol. 5, pp. 23–27, Feb. 1998.

14. L. C. Godara. "Applications of antenna arrays to mobile communications, part I: performance improvements, feasibility, and system considerations." *Proc. IEEE*, vol. 85, pp. 1031–1060, July 1997.

15. J. H. Winters. "On the capacity of radio communication systems with diversity in a Rayleigh fading environment." *IEEE JSAC*, vol. SAC-5, pp. 871–898, June 1987.

16. G. J. Foschini. "Layered space-time architecture for wireless communications in a fading environment when using multiple antennas." *Bell Labs Tech. J.*, vol. 1, pp. 41–59, Autumn 1996.

17. I. E. Telatar. "Capacity of multi-antenna Gaussian channels." *Eur. Trans. Telecomm.*, *ETT*, vol. 10, pp. 585–595, Nov./Dec. 1999.

18. S. M. Alamouti. "A simple transmit diversity technique for wireless communications." *IEEE JSAC*, vol. 16, pp. 1451–1458, Oct. 1998.

19. J. Foschini and M. J. Gans. "On the limits of wireless communications in a fading environment when using multiple antennas." *Wireless Pers. Commun.*, vol. 6, pp. 315–335, Mar. 1998.

20. P. W. Wolniansky, G. J. Foschini, G. D. Golden, and R. A. Valenzuela. "V-BLAST: an architecture for realizing very high data rates over the rich-scattering wireless channel." In *Proceedings of the URSI International Symposium on Signals, Systems, and Electronics (ISSSE 98)*, Pisa, Italy, pp. 295–300, Sep.–Oct. 1998.

21. M. Sellathurai and S. Haykin. "Turbo-BLAST: performance evaluation in correlated Rayleigh-fading environment." *IEEE JSAC*, vol. 21, pp. 340–349, Apr. 2003.

22. A. Goldsmith, S. A. Jafar, N. Jindal, and S. Vishwanath. "Capacity Limits of MIMO Channels." *IEEE JSAC*, vol. 21, pp. 684–702, June 2003.

23. H. R. Stuart. "Dispersive multiplexing in multimode optical fiber." *Science*, vol. 289, pp. 281–283, July 2000.

24. D. N. C. Tse, P. Viswanath, and L. Zheng. "Diversity-multiplexing tradeoff in multiple-access channels." *IEEE Trans. Information Theory*, vol. 50, pp. 1859–1873, Sept. 2004.

25. A. Goldsmith. *Wireless Communications*. New York: Cambridge University Press, 2005.

26. D. Chizhik, J. Ling, P. Wolniansky, R. A. Valenzuela, N. Costa, and K. Huber. "Multiple-input, multiple-output measurements and modeling in Manhattan." *IEEE JSAC*, vol. 21, pp. 321–331, April 2003.

27. M. J. Gans, N. Amitay, Y. S. Yeh, H. Xu, R. A. Valenzuela, T. Sizer, R. Storz, D. Taylor, W. M. MacDonald, C. Tran, and A. Adamiecki. "BLAST system capacity measurements at 2.44 GHz in suburban outdoor environments." In *Proceedings of the IEEE Vehicular Technology Conference (VTC 2001)*, Rhodes, Greece, vol. 1, pp. 288–292, May 2001.

28. S. Wyne, A. F. Molisch, P. Almers, G. Eriksson, J. Karedal, and F. Tufvesson. "Statistical evaluation of outdoor-to-indoor office MIMO measurements at 5.2 GHz." In *Proceedings of the IEEE Vehicular Technology Conference (VTC 2005)*, vol. 1, pp. 146–150, Spring 2005.

29. K. Yee. "Numerical solutions of initial boundary value problems involving Maxwell's equations in isotropic media." *IEEE Trans. Antennas Propagation*, vol. AP-14, pp. 302–307, 1966.

30. K. Shlager and J. Schneider. "A selective survey of the finite-difference time-domain literature." *IEEE Antennas Propagation Mag.*, vol. 37, pp. 33–56, 1995.

31. D.-S. Shiu, G. J. Foschini, M. J. Gans, and J. M. Kahn. "Fading correlation and its effect on the capacity of multielement antenna systems." *IEEE Trans. Commun.*, vol. 48, pp. 502–513, March 2000.

32. D. Gesbert, H. Bölcskei, D. A. Gore, and A. J. Paulraj. "Outdoor MIMO wireless channels: models and performance prediction." *IEEE Trans. Commun.*, vol. 50, pp. 1926–1934, 2002.

33. A. M. Sayeed. "Deconstructing multiantenna fading channels." *IEEE Trans. Signal Processing*, vol. 50, pp. 2563–2579, Oct. 2002.

34. J. Foerster. "Channel modeling subcommittee report, final," Working Group for Wireless Personal Area Networks (WPANs), IEEE P802.15-02/490r1-SG3a, Feb. 2003.

35. D. Chizhik, F. Rashid-Farrokhi, J. Ling, and A. Lozano. "Effect of antenna separation on the capacity of BLAST in correlated channels." *IEEE Commun. Lett.*, vol. 4, pp. 337–339, Nov. 2000.

36. K. Yu, M. Bengtsson, B. Ottersten, D. McNamara, P. Karlsson, and M. Beach. "A wideband statistical model for NLOS indoor MIMO channels." In *Proceedings of the Vehicular Technologies Conference (VTC 2002)*, vol. 1, pp. 370–374, May 2002.

37. W. Weichselberger. "A novel stochastic MIMO channel model and its physical interpretation." COST, Paris, France, Deliverable COST 273 TD(03)144, May 20–23 2003.

38. D. Gesbert, M. Shafi, D. Shiu, P. J. Smith, and A. Naguib. "From theory to practice: an overview of MIMO space-time coded wireless systems." *IEEE JSAC*, vol. 21, pp. 281–302, April 2003.

39. T. Kailath. "Time-variant communication channels." *IEEE Trans. Information Theory*, vol. 9, pp. 233–237, Oct. 1963.

40. P. A. Bello. "Characterization of randomly time-variant linear channels." *IEEE Trans. Commun. Systems*, vol. COM-11, pp. 360–393, Dec. 1963.

41. D. C. Cox. "Delay Doppler characteristics of multipath propagation at 910 MHz in suburban mobile radio environment." *IEEE Trans. Antennas Propagation*, vol. 20, pp. 625–635, Sept. 1972.

42. S. Howard. "Delay spread, PDP, and polarization measurements." Technical Report SCM-044-QCOM-PDPPolMeas. QUALCOMM Inc., San Diego, CA, August 2002.

43. C. L. Hong. "An improved post-processing method and an analysis of wideband channel sounding results." Technical Report CUED/F-INFENG/TR.477, University of Cambridge, November 2001.

44. E. S. Sousa and V. M. Jovanović. "Delay spread measurements for the digital cellular channel in Toronto." *IEEE Trans. Veh. Tech.*, vol. 43, pp. 1–11, Nov. 1994.

45. J. W. Wallace and M. A. Jensen. "Experimental characterization of the MIMO wireless channel." In *Proceedings of the IEEE Antennas and Propagation Society International Symposium*, vol. 3, pp. 92–95, July 2001.

46. N. Skentos, A. G. Kanatas, G. Pantos, and P. Constantinou. "Capacity results from short range fixed MIMO measurements at 5.2 GHz in urban propagation environment." In *Proceedings of the International Conference on Communications (ICC 2004)*, Paris, vol. 5, pp. 3020–3024, June 2004.

47. H. Sampath, S. Talwar, J. Tellado, V. Erceg, and A. Paulraj. "A fourth-generation MIMO-OFDM broadband wireless system: design, performance, and field trial results." *IEEE Commun. Mag.*, vol. 40, pp. 143–149, 2002.

48. A. Adjoudani, E. C. Beck, A. P. Burg, G. M. Djuknic, T. G. Gvoth, D. Haessig, S. Manji, M. A. Milbrodt, M. Rupp, D. Samardzija, A. B. Siegel, T. Sizer, C. Tran, S. Walker, S. A. Wilkus, and P. W. Wolniansky. "Prototype experience for MIMO BLAST over third-generation wireless system." *IEEE JSAC*, vol. 21, pp. 440–451, April 2003.

49. M. D. Batariere, J. F. Kepler, T. P. Krauss, S. Mukthavaram, J. W. Porter, and F. W. Vook. "An experimental OFDM system for Broadband mobile communications." In *Proceedings of the Vehicular Technology Conference (VTC 2001)*, Atlantic City, NJ, vol. 4, pp. 1947–1951, Oct. 2001.

50. H. Özcelik, M. Herdin, W. Weicheselberger, J. Wallace, and E. Bonek. "Deficiencies of the Kronecker MIMO radio channel model." *IEEE Electronics Lett.*, vol. 39, pp. 1209–1210, August 2003.

51. W. Weichselberger. "Spatial structure of multiple antenna radio channels." Institut für Nachrichtentechnik und Hochfrequenztechnik. Technische Universität Wien, PhD Thesis, 2003.

52. W. Tuttlebee. *Software Defined Radio: Baseband Technology for 3G Handsets and Basestations*. New York: John Wiley & Sons, 2004.

53. J. Mitola. "Software radios: survey, critical evaluation and future directions." *IEEE Aerospace and Electronic Systems Magazine*, vol. 8, no. 4, pp. 25–36, April 1993.

54. J. Mitola. "The software radio architechture." *IEEE Commun. Mag.*, vol. 33, pp. 26–38, May 1995.

55. The Software Defined Radio Forum homepage. Available at: http://www.sdrforum.org/pages/aboutTheForum/aboutTheForum.asp.

56. The Third Generation Partnership Project. Available at: http://www.3gpp.org/About/about.htm.

57. W. Weichselberger, M. Herdin, H. Özcelik, and E. Bonek. "A stochastic MIMO channel model with joint correlation of both link ends." *IEEE Trans. Wireless Commun.*, vol. 5, pp. 90–100, Jan. 2006.

58. T. S. Rappaport. *Wireless Communications: Principles and Practice*. Upper Saddle River, NJ: Prentice Hall, 1996.

59. W. C. Y. Lee. *Mobile Cellular Telecommunications Systems*. New York: McGraw-Hill, 1989.

60. D. D. Durgin. *Space-Time Wireless Channels*. Upper Saddle River, NJ: Prentice Hall, 2003.

61. J. G. Proakis and D. G. Manolakis. *Digital Signal Processing: Principles, Algorithms, and Applications*, 3rd ed. Upper Saddle River, NJ: Prentice Hall, 1996.

62. L. Zheng and D. N. C. Tse. "Diversity and multiplexing: a fundamental tradeoff in multiple-antenna channels." *IEEE Trans. Information Theory*, vol. 49, pp. 1073–1096, May 2003.

63. G. Bauch and N. Al-Dhahir. "Reduced-complexity space-time turbo-equalization for frequency-selective MIMO channels." *IEEE Trans. Wireless Commun.*, vol. 1, pp. 819–828, Oct. 2002.

64. W. Weichselberger. "On the decomposition of the MIMO channel correlation tensor." In *ITG Workshop on Smart Antennas*, pp. 268–273, March 2004.

65. R. Janaswamy. *Radiowave Propagation and Smart Antennas for Wireless Communications*. Boston, MA: Kluwer Academic Publishers, 2000.

66. J. Litva and T. K.-Y. Lo. *Digital Beamforming in Wireless Communications*. Norwood, MA: Artech House, 1996.

67. H. Krim and M. Viberg. "Two decades of array signal processing research: the parametric approach." *IEEE Signal Processing Mag.*, vol. 13, pp. 67–94, July 1996.

68. A. V. Oppenheim, A. S. Willsky, and S. H. Nawab. *Signals & Systems*, 2nd ed. Upper Saddle River, NJ: Prentice Hall, 1996.

69. L. A. Zadeh. "Time-varying networks, I." *Proc. IRE*, vol. 49, pp. 1488–1502, Oct. 1961.

70. T. Kailath. "Measurements on time-variant communications channels." *IRE Trans. Information Theory*, vol. 8, pp. 229–236, Sept. 1962.

71. M. A. Jensen and J. W. Wallace. "MIMO wireless channel modeling and experimental characterization." In *Space-Time Processing for MIMO Communications*, A. B. Gershman and N. D. Sidiropoulos, Eds., John Wiley & Sons, Inc., NJ, 2005, pp. 1–39.

72. K. I. Pedersen, J. B. Andersen, J. P. Kermoal, and P. Mogensen. "A stochastic multiple-input-multiple-output radio channel model for evaluation of space-time coding algorithms." In *Proceedings of the IEEE Vehicular Technology Conference (VTC 2000)*, vol. 1, pp. 893–897, Fall 2000.

73. K. Yu. *Multiple-Input Multiple-Output Radio Propagation Channels: Characteristics and Models*. Stockholm Sweden: KTH University, 2005.

74. H. Özcelik, N. Czink, and E. Bonek. "What makes a good MIMO channel model?" In *Proceedings of the Vehicular Technology Conference (VTC 2005)*, vol. 1, pp. 156–160, May–June 2005.

75. L. D. Lathauwer. "A multilinear singular value decomposition." *Society for Industrial and Applied Mathematics J. Matrix Anal. Appl.*, vol. 21, pp. 1253, 1278, March–May 2000.

76. K. Yu, M. Bengtsson, B. Ottersten, D. McNamara, P. Karlsson, and M. Beach. "Second order statistics of NLOS indoor MIMO channels based on 5.2 GHz measurements." In *Proceedings of the IEEE Global Telecommunications Conference (GLOBECOM 2001)*, San Antonio, TX, vol. 1, pp. 156–160, Nov. 2001.

77. N. Costa and S. Haykin. "A novel wideband channel model and experimental validation." *Trans. Antennas Propagation*, vol. 56, pp. 550–562, Feb. 2008.

78. M. Steinbauer, A. Molisch, and E. Bonek. "The double-directional radio channel." *IEEE Antennas Propagation Mag.*, vol. 43, pp. 51–63, Aug. 2001.

79. L. Hentilä, P. Kyösti, M. Käske, M. Narandzic, and M. Alatossava. "MATLAB implementation of the WINNER Phase II Channel Model ver1.1." Available at: http://projects. celtic-initiative.org/winner+/phase_2_model.html.

80. N. Czink. "The random-cluster model – a stochastic MIMO channel model for broadband wireless communication systems of the 3rd generation and beyond." In *Institut für Nachrichtentechnik und Hochfrequenztechnik*. Vienna: Technische Universität Wien, Available at: publik.tuwien.ac.at/files/PubDat_112121.pdf.

81. G. L. Turin, F. D. Clapp, T. L. Johnsoton, S. B. Fine, and D. Lavry. "A statistical model of urban multipath propagation." *IEEE Trans. Veh. Tech.*, vol. VT-21, pp. 1–9, Feb. 1972.

82. A. A. M. Saleh and R. A. Valenzuela. "A statistical model for indoor multipath propagation." *IEEE JSAC*, vol. SAC-5, pp. 128–137, Feb. 1987.

83. A. F. Molisch, H. Asplund, R. Heddergott, M. Steinbauer, and T. Zwick. "The COST 259 directional channel model - I. overview and methodology." *IEEE Trans. Wireless Commun.*, vol. 5, pp. 3421–3433, 2006.

84. H. Asplund, A. A. Glazunov, A. F. Molisch, K. I. Pedersen, and M. Steinbauer. "The COST 259 directional channel model - II. macrocells." *IEEE Trans. Wireless Commun.*, vol. 5, pp. 3434–3450, 2006.

85. J. Laurila, K. Hugl, M. Toeltsch, E. Bonek, K. Kalliola, and P. Vainikainen. "Directional wideband 3-D measurements of mobile radio channel in urban environment." *COST 259, TD(99)092*, Leidschendam, The Netherlands, Sept. 1999.

86. C.-C. Chong, C.-M. Tan, D. Laurenson, S. McLaughlin, M. Beach, and A. Nix. "A new statistical wideband spatio-temporal channel model for 5-GHz band WLAN systems." *IEEE JSAC*, vol. 21, pp. 130–150, Feb. 2003.

87. K. Yu, Q. Li, D. Cheung, and C. Prettie. "On the tap and cluster angular spreads of indoor WLAN channels." In *Proceedings of the IEEE Vehicular Technology Conference Spring 2004*, Milano, Italy, vol. 1, pp. 17–19, May 2004.

88. N. Czink, X. Yin, H. Özcelik, M. Herdin, E. Bonek, and B. Fleury. "Cluster characteristics in a MIMO indoor propagation environment." *IEEE Trans. Wireless Commun.*, vol. 6, pp. 1465–1475, Apr. 2007.

89. J. Salo, J. Salmi, N. Czink, and P. Vainikainen. "Automatic clustering of nonstationary MIMO channel parameter estimates." In *Proceedings of the 2nd International Conference on Telecommunications (ITC 2005)*, Cape Town, South Africa, May 2005.

90. 3GPP. "Spatial channel model for multiple input multiple output (MIMO) simulations (3GPP TR 25.996), v.6.1.0." Available at: www.3gpp.org.

91. L. M. Correia. *Mobile Broadband Multimedia Networks: Techniques, Models and Tools for 4G*. London: Elsevier, 2006.

92. J. Foerster. "IEEE P802.15 Working Group for Wireless Personal Area Networks (WPANs): Channel modeling sub-committee report final." Available at: www.ieee802. org/15/pub/2003/Mar03/02490r1P802-15_SG3a-Channel-Modeling-Subcommittee-Report-Final.zip.

93. N. Czink, E. Bonek, X. Yin, and B. Fleury. "Cluster angular spreads in a MIMO indoor propagation environment." In *Proceedings of the IEEE 16th Personal International Symposium on Indoor and Mobile Radio Communications (PIMRC 2005)*, Berlin, Germany, vol. 1, pp. 664–668, Sept. 2005.

94. Q. H. Spencer, B. D. Jeffs, M. A. Jensen, and A. L. Swindlehurst. "Modeling the statistical time and angle of arrival characteristics of an indoor multipath channel." *IEEE J. Selected Areas Commun.*, vol. 18, pp. 347–360, Mar. 2000.

95. B. Fleury, P. Jourdan, and A. Stucki. "High-resolution channel parameter estimation for MIMO applications using the SAGE algorithm." In *2002 International Zurich Seminar on Broadband Communications*, Zurich, Switzerland, pp. 30-1–30-9, 2002.

96. B. H. Fleury, M. Tschudin, R. Heddergott, D. Dahlhaus, and K. I. Pedersen. "Channel parameter estimation in mobile radio environments using the SAGE algorithm." *IEEE JSAC*, vol. 17, pp. 434–450, Mar. 1999.

97. H. Hofstetter. "The COST 273 MIMO channel model implementation." Vienna: Telecommunications Research Center Vienna. Available at: http://userver.ftw.at/cost273/index.htm.

98. A. F. Molisch. "A generic channel model for MIMO wireless propagation channels in macro- and microcells." *IEEE Trans. Signal Processing*, vol. 52, pp. 61–71, Jan. 2004.

99. T. Klingenbrunn and P. Mogensen. "Modelling cross-correlated shadowing in network simulations." In *Proceedings of the Vehicular Technology Conference (VTC 1999)*, Amsterdam, vol. 3, pp. 1407–1411, Sept. 1999.

100. I. Ihler. "Kernel density estimation toolbox for Matlab." Available at: http://ttic.uchicago.edu/~ihler/code/.

101. S. Haykin. *Adaptive Filter Theory*, 4th ed. Upper Saddle River, NJ: Prentice Hall, 2002.

102. S. M. Kay. *Fundamentals of Statistical Processing, Estimation Theory*. Upper Saddle River, NJ: Prentice Hall, 1993.

103. N. Czink, P. Cera, J. Salo, E. Bonek, J.-P. Nuutinen, and J. Ylitalo. "Improving clustering performance by using the multi-path component distance." *IEEE Electronic Lett.*, vol. 42, pp. 44–45, Jan 2006.

104. N. Czink, E. Bonek, J. Ylitalo, and T. Zemen. "Measurement-based time-variant MIMO channel modelling using clusters." In *Proceedings of the XXIX General Assembly of the International Union of Radio Science (URSI)*, Chicago, IL, August 2008.

105. J. Salo, G. D. Galdo, J. Salmi, P. Kyösti, M. Milojevic, D. Laselva, and C. Schneider. "MATLAB implementation of the 3GPP spatial channel model (3GPP TR 25.996)." Available at: http://www.mathworks.com/matlabcentral/fileexchange/loadFile.do?objectId=20911&objectType=File.

106. S. Haykin. *Communication Systems*, 4th ed. New York: John Wiley & Sons, 2001.

107. P. Vizmuller. *RF Design Guide: Systems, Circuits, And Equations*. Norwood, MA: Artech House Inc., 1995.

108. P. G. Flikkema and S. G. Johnson. "A comparison of time- and frequency-domain wireless channel sounding techniques." In *Proceedings of the IEEE Southeastcon '96*, Tampa, FL, pp. 488–491, April 1996.

109. J. G. Proakis. *Digital Communications*, 3rd ed. Boston, MA: McGraw-Hill, 1995.

110. "M-sequence generator." Matlab Central File Exchange. Available at: http://www.mathworks.com/matlabcentral/fileexchange/loadFile.do?objectId=990&objectType=file.

111. P. C. Fannin, A. Molina, S. S. Swords, and P. J. Cullen. "Digital signal processing techniques applied to mobile radio channel sounding." *IEE Proc.-F*, vol. 138, pp. 502–508, Oct. 1991.

112. Wikipedia. "Maximum length sequence." Available at: http://en.wikipedia.org/wiki/Maximum_length_sequence.

113. G. Matz, A. F. Molisch, M. Steinbauer, F. Hlawatsch, I. Gaspard, and H. Artes. "Bounds on the systematic measurement errors of channel sounders for time-varying mobile radio channels." In *Proceedings of the Vehicular Technology Conference (VTC 1999)*, Amsterdam, vol. 3, pp. 1465–1470, Sept. 1999.

114. G. Matz, A. R. Molisch, F. Hlawatsch, M. Steinbauer, and I. Gaspard. "On the systematic measurement errors of correlative mobile radio channel sounders." *IEEE Trans. Commun.*, vol. 50, pp. 808–821, May 2002.

115. MEDAV GmbH Homepage. Available at: http://www.medav.de/.

116. J. W. Wallace. *SDRC Data Collection Campaign and Channel Modeling.* Provo City, UT: Brigham Young University, 2004.

117. Symmetricom Home Page. Available at: http://symmetricom.com/.

118. H. Meyr, M. Meclaey, and S. A. Fechtel. *Digital Communication Receivers: Synchronization, Channel Estimation, and Signal Processing.* New York: John Wiley & Sons, 1998.

119. M. Paier, A. Molisch, and E. Bonek. "Training-sequence-based determination of optimum sampling time in unequalized TDMA mobile radio systems." *IEEE Trans. Veh. Tech.,* vol. 49, pp. 1465–1470, July 2000.

120. G. D. Golden, G. J. Foschini, P. W. Wolniansky, and R. A. Valenzuela. "V-BLAST: A high capacity space-time architechture for the rich-scattering wireless channel." In *Proceedings of the International Symposium on Advanced Radio Technologies,* Boulder, CO. Available at: http://www1.bell-labs.com/project/blast/, Sept. 1998.

121. G. D. Golden, C. J. Foschini, R. A. Valenzuela, and P. W. Wolniansky. "Detection algorithm and initial laboratory results using V-BLAST space-time communication architecture." *IEEE Electronic Lett.,* vol. 35, pp. 14–16, Jan. 1999.

122. "BLAST: Bell Labs Layered Space-Time, an architecture for realizing very high data rates over fading wireless channels." Holmdel, NJ.: Available at: http://www.bell-labs.com/project/blast/, 2000.

123. J. S. Aron. *Measurement System and Campaign for Characterizing of Theoretical Capacity and Cross-correlation of Multiple-Input Multiple Output Indoor Wireless Channels.* Blacksburg, VA: Viginia Polytechnic Institute and State University, 2002.

124. Electrobit PropSim homepage. Available at: http://www.propsim.com/.

125. BYU Wireless Communications Research Homepage. Available at: http://www.ece.byu.edu/wireless/.

126. J. W. Wallace, M. A. Jensen, A. L. Swindlehurst, and B. D. Jeffs. "Experimental characterization of the MIMO wireless channel: data acquisition and analysis." *IEEE Trans. Wireless Commun.,* vol. 2, no. 2, pp. 335–343, March 2003.

127. M. T. Ivrlač and J. A. Nossek. "Quantifying diversity and correlation in Rayleigh fading MIMO communication systems." In *Proceedings of the 3rd IEEE International Symposium on Signal Processing and Information Technology (ISSPIT 2003),* Darmstadt, Germany, pp. 158–161, December 2003.

128. C. D. Meyer. *Matrix Analysis and Applied Linear Algebra.* Philadelphia: SIAM, 2000.

129. J. W. Demmel. "The probability that a numerical analysis problem is difficult." *Math. Computat.,* vol. 50, pp. 449–480, 1988.

130. P. Kyösti, D. Laselva, L. Hentilä, and T. Jämsä. "Validating IST-WINNER indoor MIMO radio channel model." In *Proceedings of the IST Mobile and Wireless Summit 2006,* Mykonos, Greece, June 2006.

131. N. Czink, G. D. Galdo, X. Yin, E. Bonek, and J. Ylitalo. "A novel environment characterization metric for clustered MIMO channels used to validate a SAGE parameter estimator." *Wireless Pers. Commun.,* vol. 46, pp. 83–98, July 2008.

132. N. Czink, T. Zemen, J.-P. Nuutinen, J. Ylitalo, and E. Bonek. "A time-variant MIMO channel model directly parameterised from measurements." *EURASIP Journal on Wireless Communications Networking, Special Issue on Advances in Propagation Modeling for Wireless Systems,* vol. 2009, Article ID 687238, 2009.

133. L. Hentilä, P. Kyösti, J. Ylitalo, X. Zhao, J. Meinilä, and J.-P. Nuutinen. "Experimental characterization of multi-dimensional parameters at 2.45 and 5.25 GHz indoor channels." In *Proceedings of the Wireless Personal Multimedia Communications*, Aalborg, Denmark, Sept. 2005.

134. H. MacLeod, C. Loadman, and Z. Chen. "Experimental studies of the 2.4-GHz ISM wireless indoor channel." In *Proceedings CNSR 2005*, 2005, pp. 63–88.

135. J. Cullen, P. C. Fannin, and A. Molina. "Wide-band measurement and analysis techniques for the mobile radio channel." *IEEE Trans. Veh. Tech.*, vol. 42, pp. 589–603, Nov. 1993.

136. L. D. Lathauwer. *Signal Processing Based on Multilinear Algebra*. Leuven, Belgium: Katholieke Universiteit Leuven, 1997.

137. J. H. Heinbockel. *Introduction to Tensor Calculus and Continuum Mechanics*. Trafford, Victoria, BC, 2006.

138. M. Alex, O. Vasilescu, and D. Terzopoulos. "Multilinear analysis of image ensembles: TensorFaces." In *Proceedings of the European Conference on Computer Vision*, Copenhagen, Denmark, pp. 447–460, May 2002.

139. Mathworks. atan2 (Matlab Functions). Available at: http://www.mathworks.com/access/helpdesk_r13/help/techdoc/ref/atan2.html.

INDEX

Multiple-Input, Multiple-Output Channel Models: Theory and Practice. By Nelson Costa and Simon Haykin
Copyright © 2010 John Wiley & Sons, Inc.